ENGLISH for ARTIFICIAL INTELLIGENCE
Reading and Writing
人工智能英语读写教程

主　编　何　芳
副主编　郑　涛　邓　静
编　者　孟庆丰　刘晓玲
　　　　封国华　何　芸
　　　　刘　凤　李新国

清华大学出版社
北　京

内 容 简 介

本教材共八个单元，立足人工智能技术的前沿动态，从专业角度对人工智能的发展概况、融合与应用以及未来前景等内容进行介绍。本教材以真实性为原则，选用专业性强、时效性高的阅读素材，助力学生的专业学习；单元结构层次清晰，主要由阅读文章、写作技巧、翻译技巧和课外拓展四个板块构成，旨在拓宽学生的知识面并提高他们的科技英语应用能力；课后练习形式多样，针对基础的语言技能展开训练，针对性和实用性强。

本教材既可作为高校计算机、机器人等相关专业学生的专业英语教材，也可供广大科技人员和科技英语爱好者阅读参考。

图书在版编目（CIP）数据

人工智能英语读写教程 / 何芳主编. —北京：清华大学出版社，2021.7
高校专门用途英语（ESP）系列教材
ISBN 978-7-302-57807-9

Ⅰ. ①人… Ⅱ. ①何… Ⅲ. ①人工智能—英语—阅读教学—高等学校—教材 ②人工智能—英语—写作—高等学校—教材 Ⅳ. ①TP18

中国版本图书馆CIP数据核字（2021）第055383号

责任编辑： 刘 艳
封面设计： 子 一
责任校对： 王凤芝
责任印制： 沈 露

出版发行： 清华大学出版社
网 址： http:// www. tup. com. cn，http:// www. wqbook. com
地 址： 北京清华大学学研大厦A座 **邮 编：** 100084
社 总 机： 010-62770175 **邮 购：** 010-62786544
投稿与读者服务： 010-62776969, c-service@tup.tsinghua.edu.cn
质量反馈： 010-62772015, zhiliang@tup.tsinghua.edu.cn
印 装 者： 三河市龙大印务有限公司
经 销： 全国新华书店
开 本： 185mm × 260mm **印 张：** 15.5 **字 数：** 339 千字
版 次： 2021 年 7 月第 1 版 **印 次：** 2021 年 7 月第 1 次印刷
定 价： 69.00 元

产品编号：091770-01

前　言

当前，全社会都在关注人工智能。为了抢抓人工智能发展的重大战略机遇，构筑我国人工智能发展的先发优势，加快建设创新型国家和世界科技强国，国家对具有较高专门应用水平的专业人才培养提出了新的要求，并出台了《新一代人工智能发展规划》。在这样的背景下，北京联合大学于2016年在全国率先成立机器人学院，着重发展智能机器人技术，以期为国家人工智能产业的建设与发展提供高素质复合型应用人才，特别是接轨国际的高素质人才。

《人工智能英语读写教程》是在充分调研学科需求和国内外相关教材的基础上，在目前国内市场上缺乏同类英语教材的背景下做出的一次积极尝试。本教材注重为学生提供有针对性的专业英语学习素材，模拟行业英语应用的真实场景，促进学生在专业领域和行业方向的语言学习，从而培养学生特定的语言应用能力。

本教材将人工智能领域所需的语言知识和技能需求密切结合，从专业角度对人工智能的发展概况、人工智能与各领域的融合和应用以及人工智能的未来前景等方面做了比较全面的介绍。本教材共八个单元，包括人工智能概述、人工智能与机器学习、人工智能与大数据、人工智能与云计算、人工智能与物联网、人工智能的应用和人工智能的未来前景等方面内容。每单元开篇明确本单元的学习目标；之后围绕单元主题介绍Text A、Text B与Text C三篇文章，并附有Notes、Words and Expressions、Useful Terms、Exercises等板块；最后设有Writing Skills、Translating Skills和Workshop板块，培养学生的科技英语写作、翻译以及综合应用能力。

本教材具有以下特色：

1. 以真实性为原则

所有阅读材料均选自国外权威杂志和知名网站，不仅素材真实、语言地道，而且具有时效性。单元内容涉及真实的公司和人物介绍，以及真实场景下的案例分析，具有较强的实用性和专业性，帮助学生提前熟悉并逐渐适应英语语言环境下人工智能相关领域的工作内容。

2. 内容丰富且具有深度

每个单元从阅读、写作和翻译方面着手，注重对学生学习、思考和应用能力的启发和培养。其中，Text A 和 Text B 两篇文章不仅与主题高度相关，而且侧重从不同方面培养学生的专业素养与语言技能，分别适用于精读和泛读，每篇文章长度约 800~1000 词；Text C 是为学生准备的自学材料，不仅标注了重点词汇，还提供了必要的汉语释义，每篇文章长度约 900~1200 词。

另外，每单元还提供了系统的科技英语写作指导、翻译技巧讲解以及与人工智能技术相关的新闻报道等，帮助学生学习理论知识之余，在学术英语、相关背景知识和社会话题探讨方面得到充分的语言输入。

3. 设有特色练习

Workshop 板块以本单元的语言应用能力为目标，以任务驱动学习为设计原理，以口头演讲、合作研究和案例分析等方式实践运用单元重点知识和语言技能，充分激发学生的主动意识和进取精神，从而促进学生的有效学习。

4. 配套资源丰富

本教材配有丰富的教学资源，包括电子版的教学参考书、PPT 课件、精品慕课等，供老师和学生参考使用。相关资源可通过点击 ftp://ftp:tup.tsinghua.edu.cn 下载使用。

本教材既可以作为高校本科计算机、机器人等相关专业学生的专业英语教材，也可以作为全国各类人工智能行业机构和组织的专业培训教材，还可以作为从事人工智能领域相关工作的企业管理者和技术人员的学习材料。

本教材的编写团队来自北京联合大学，何芳担任主编，郑涛、邓静担任副主编，刘晓玲担任审校。郑涛、邓静各编写 1.5 个单元，孟庆丰、封国华、何芸、刘凤和李新国各编写 1 个单元。由于编者水平有限，书中难免存在疏漏之处，请广大读者不吝指正。

编者

2021 年 5 月

Contents

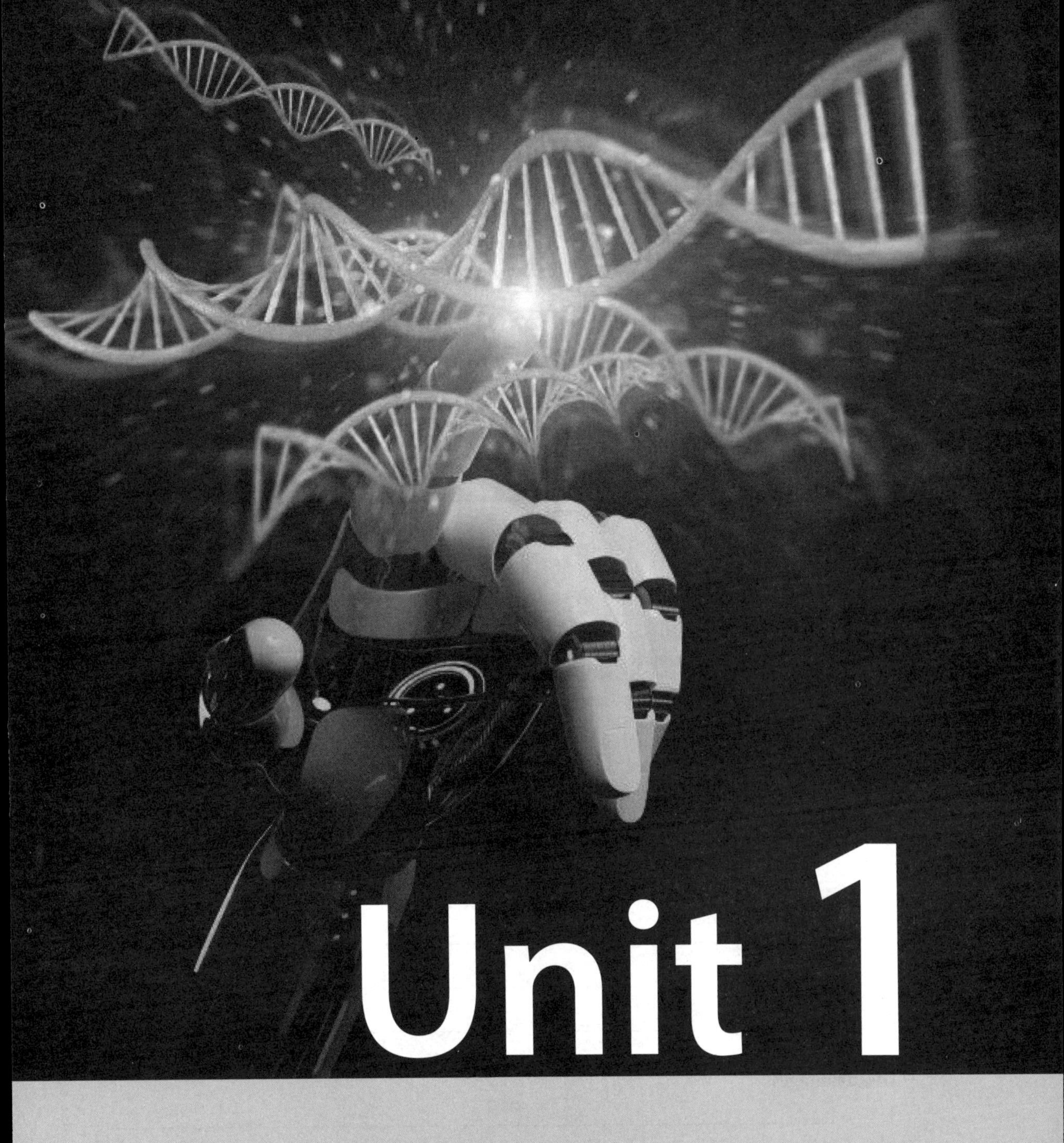

Unit 1

An Overview of Artificial Intelligence

Learning Objectives

In this unit, you will learn:

- an overview of artificial intelligence;
- the history of artificial intelligence;
- the current state of artificial intelligence;
- writing skills—different ways of defining;
- translating skills—translation of the passive voice.

Lead-in

I. Discuss the following questions with your partners.

1. How much do you know about artificial intelligence?
2. What are the goals of artificial intelligence?
3. In what ways can artificial intelligence influence people's life and work?

II. Work in pairs to discuss the advantages and disadvantages of artificial intelligence.

Advantages:

1. ______________________________
2. ______________________________
3. ______________________________

Disadvantages:

1. ______________________________
2. ______________________________
3. ______________________________

An Overview of Artificial Intelligence

1 Since the invention of computers or machines, their capability to perform various tasks went on growing **exponentially**. The power of computer systems **in terms of** their **diverse** working **domains**, their increasing speed, and their reducing size **with respect to** time has been greatly developed.

2 A branch of computer science named artificial intelligence (AI) pursues creating computers or machines as intelligent as human beings.

What Is Artificial Intelligence?

3 According to the father of artificial intelligence, John McCarthy, it is "the science and engineering of making intelligent machines, especially intelligent computer programs".

4 Artificial intelligence is a way of making a computer, a computer-controlled robot, or a piece of software that can think as intelligently as humans.

5 AI is accomplished by studying how human brain thinks, and how humans learn, decide, and work while trying to solve a problem, and then using the outcomes of this study as a basis of developing intelligent software and systems.

6 It is one thing to explain why AI is exciting; however, it is another to explain what AI is. We could just say, "Well, it has to do with smart programs, so let's get on and write some." But the history of science shows that with the right goal, AI is developing in the right direction. Early **alchemists**, looking for a **potion** for eternal life and a method to **turn** lead **into** gold, **were** probably **off on the wrong foot**. The scientific method could emerge and productive science could take place when the aim changed to find **explicit** theories that gave accurate predictions of the **terrestrial** world, in the same way that early astronomy predicted the apparent motions of the stars and planets.

Philosophy of Artificial Intelligence

7 The study of the power of the computer system arouses human's curiosity: Can a machine think and behave like humans do?

8 Thus, the development of AI started with the intention of creating similar intelligence in machines that we find and regard high in humans.

Goals of Artificial Intelligence

9 The goals of developing AI are:

- To create expert systems—the systems which display intelligent behavior, learn, demonstrate, explain, and advise its users.
- To implement human intelligence in machines—creating systems that can think and act as intelligently as humans.

What Contributes to Artificial Intelligence?

10 **Disciplines** involved in artificial intelligence include computer science, biology, psychology, linguistics, mathematics, and engineering. A major thrust of AI is in the development of computer functions associated with human intelligence, such as reasoning, learning, and problem-solving.

Applications of Artificial Intelligence

11 AI has been extensively used in various fields such as:

- Gaming—AI plays a crucial role in strategic games such as chess, poker, tic-tac-toe, etc., where machines can think of a large number of possible positions based on **heuristic** knowledge.
- Natural language processing—it is possible to interact with the computer that understands natural language spoken by humans.
- Expert systems—there are some applications which integrate machine, software, and special information to impart reasoning and advising. They provide explanation and professional advice for the users.
- Vision systems—these systems distinguish and interpret visual input on the computer. For example, a spying aeroplane takes photographs and the photographs can be used to analyze spatial information or map of the areas. Clinical expert systems can help doctors better **diagnose** and treat patients. With the help of computer systems, police can recognize the appearance of criminals with the stored portrait made by **forensic** artist.
- Speech recognition—some intelligent systems are capable of hearing and comprehending the language a human talks to them. They can deal with different accents, slang words, the background noise, even the change in human's sound due to cold, etc.
- Handwriting recognition—all texts written on paper by a pen or on screen by a **stylus** are readable via the handwriting recognition system, which can "read" the shapes of the letters and change them into editable text.
- Intelligent robots—robots are capable of performing the tasks assigned by a human. They have sensors to collect and analyze physical data from the real world such as light, heat, temperature, movement, sound, bump, and pressure. They are called intelligent robots just because they have efficient processors, multiple sensors, and huge memory. Additionally,

they can improve themselves by learning from their mistakes and adapting to the new environment.

12 AI addresses one of the ultimate puzzles. How can a slow, tiny brain, biological or electronic, **perceive**, understand, predict, and manipulate a world far larger and more complicated than itself? How do we go about making something with those properties? It is not easy to answer these questions, but unlike the search for faster-than-light travel or an **antigravity** device, the researchers in AI have solid evidence that the quest is possible. All the researchers have to do is to look in the mirror to see an example of an intelligent system.

13 AI is one of the freshest disciplines. It was formally initiated in 1956, when the name was coined, although at that point work had been **underway** for about five years. Along with modern **genetics**, it is regularly cited as the "field I would most like to be in" by scientists in other disciplines. A student in physics might have good reasons to feel that all the great ideas have already been taken by Galileo, Newton, Einstein, and other outstanding scientists, and that it takes many years of study before one can contribute new ideas. AI, on the other hand, still has openings for a full-time Einstein.

Notes

John McCarthy an American computer scientist and cognitive scientist. McCarthy was one of the founders of the discipline of artificial intelligence. He coined the term "artificial intelligence", developed the Lisp programming language family, significantly influenced the design of the ALGOL programming language, popularized timesharing, and was very influential in the early development of AI.

tic-tac-toe also known as noughts and crosses, or Os and Xs, a paper-and-pencil game for two players, O and X, who take turns marking the spaces in a 3 × 3 grid. The player who succeeds in placing three of his marks in a horizontal, vertical, or diagonal row wins the game.

Galileo an Italian astronomer, physicist, and engineer, who was sometimes described as a polymath. Galileo has been called the father of observational astronomy, the father of modern physics, the father of the scientific method, and the father of modern science.

Newton an English mathematician, physicist, astronomer, theologian, and author (described in his own day as a "natural philosopher"), who is widely recognized as one of the most influential scientists of all time, and a key figure in the Scientific Revolution (1500–1750).

Einstein a German-born theoretical physicist who developed the theory of relativity, one of the two pillars of modern physics (alongside quantum mechanics). His work is also known for its influence on the philosophy of science. He is best known to the general public for his mass-energy equivalence formula $E = mc^2$, which has been dubbed "the world's most famous equation". He received the 1921 Nobel Prize in Physics "for his services to theoretical physics, and especially for his discovery of the law of the photoelectric effect", a pivotal step in the development of quantum theory.

Words and Expressions

exponentially /ˌekspəˈnenʃəli/ *adv.* in a way that becomes faster and faster 以指数方式

in terms of with regard to 从……的角度；用……来表示

diverse /daɪˈvɜːs/ *adj.* very different from each other and of various kinds 不同的；变化多的

domain /dəʊˈmeɪn/ *n.* an area of knowledge or activity, especially one that somebody is responsible for （知识、活动的）领域，范围

with respect to in relation to 关于；谈到

alchemist /ˈælkəmɪst/ *n.* one who is versed in the practice of alchemy and who seeks an elixir of life and a panacea 炼金术士

potion /ˈpəʊʃn/ *n.* a drink of medicine or poison; a liquid with magic power 药饮；毒液；魔水

turn…into to make somebody/something become somebody/something else 把……转变成……

be off on the wrong foot to begin a relationship or project badly 一开始就错了

explicit /ɪkˈsplɪsɪt/ *adj.* clear and easy to understand, so that you have no doubt what is meant 详尽的；明确的

terrestrial /təˈrestriəl/ *adj.* connected with the planet Earth 地球上的

discipline /ˈdɪsəplɪn/ *n.* a subject that people study or are taught, especially in a university; an area of knowledge （大学）学科；知识领域

heuristic /hjʊəˈrɪstɪk/ *adj.* enabling a person to discover or learn something for himself/herself 启发式的

diagnose /ˈdaɪəgnəʊz/ *v.* to say exactly what an illness or the cause of a problem is 诊断

forensic /fəˈrenzɪk/ *adj.* connected with the scientific tests used by the police when trying to solve a crime 法医的

stylus /ˈstaɪləs/ *n.* a special pen used to write text or draw an image on a

			special computer screen 触笔；指示笔
perceive	/pə'siːv/	*v.*	to notice or become aware of something 觉察，意识到
antigravity	/ˌænti'grævɪti/	*n.*	an imaginary force that works against gravity 反重力，反引力
underway	/ˌʌndə'weɪ/	*adj.*	currently in progress 进行中的
genetics	/dʒə'netɪks/	*n.*	scientific study of the ways in which different characteristics are passed from each generation of living things to the next 遗传学

Useful Terms

working domain	工作领域
spying aeroplane	侦察飞机
forensic artist	法医艺术家

Exercises

Comprehension Check

I. Answer the following questions according to the text.

1. Who is the father of artificial intelligence?
2. What is the philosophy of artificial intelligence?
3. What are the goals of artificial intelligence?
4. What is the basis of the research in artificial intelligence?
5. In what ways can artificial intelligence help the police?

II. Read the following statements carefully and decide whether they are true (T) or false (F) without turning back to check the text.

1. ______________ Artificial intelligence aims at making a computer or a piece of software that can think more intelligently than human beings.
2. ______________ The development of artificial intelligence is based on the study of the

way human beings think, learn, and work.

3. ______________ It is the goal of finding theories that can lead scientists to make breakthroughs in productive science.

4. ______________ Artificial intelligence is a science that is closely related to physics.

5. ______________ Artificial intelligence can be reliable when people need theoretical explanation or technical advice.

III. Choose the best answer to each of the following questions according to the text.

1. Artificial intelligence is the science and engineering of ______________.

   A. creating machines that can calculate faster than human beings

   B. creating systems that can think, learn, and understand as intelligently as human beings

   C. creating machines that can store larger amounts of data than human beings

   D. creating systems that can think, decide, and reason better than human beings

2. Artificial intelligence can be applied in the following fields EXCEPT ______________.

   A. playing games

   B. handling different accents and voices

   C. going into the war

   D. performing extreme tasks

3. The handwriting recognition system can read texts by ______________.

   A. analyzing the grammatical rules

   B. translating the texts

   C. scanning the key information

   D. identifying the shapes of the letters

4. It is possible for AI to become more advanced if scientists can ______________.

   A. study how computers work normally

   B. study how human brain works normally

   C. study how large amounts of data can be stored

   D. study how computing speed can be increased

5. Why does the author mention "a full-time Einstein"?

   A. Because AI is a new field in which anyone can make new discoveries and contributions.

B. Because AI is a brand-new field in which scientists have to give up their previous research.

C. Because AI is so complex that only those top scientists can do research in it.

D. Because AI is such a complicated study that one has to fully concentrate on it.

Vocabulary Building

IV. Fill in the following blanks with the words and phrases given in the box. Change the form if necessary.

diverse	pursue	explicit	prediction	function
discipline	underway	in terms of	turn…into	be associated with

1. People with a wide range of ______________ are more competitive in the job market.
2. Plans are already ______________ for the optimization of the application system.
3. Global warming ______________ people's life closely and disastrous consequences would be unavoidable if we don't do anything to cope with it.
4. No one should make a(n) ______________ about which IT product would be the most popular in the coming years.
5. All the top universities in the world are appealing to students from all over the world not only because of their excellence in teaching and researching but also because of their cultural ______________.
6. The Republic of Botswana is a small country ______________ size and population.
7. An increasing number of college students start to ______________ their own business after they graduate.
8. It took them three weeks to ______________ the old car park ______________ a playing ground for the residents nearby.
9. The equipment is now ______________ normally thanks to the maintenance of the engineers.
10. The professor has been talking ______________ about the development and application of AI.

V. Match the words in the left column with the explanations in the right column.

1. robot	A. the branch of physics that studies celestial bodies and the universe as a whole
2. prediction	B. of extreme importance

3.	astronomy	**C.**	a mechanism that can move automatically
4.	crucial	**D.**	involving several
5.	multiple	**E.**	the statement made about the future

Word Formation

VI. Fill in the following blanks with the words in capitals. Change the form if necessary. An example has been given.

e.g. *It can recognize the shape of the letters and convert it into* <u>*editable*</u> *text.* **EDIT**

1. The installation of ______________ traffic signs is expected to be effective in reducing the number of accidents in the city. **CHANGE**
2. The seats in this theater are ______________ so that the audience will feel comfortable while watching the performance on the stage. **ADJUST**
3. The mailman had trouble in sending the parcel because the address on it was not ______________. **READ**
4. It's ______________ that she was annoyed by such an offensive behavior. **UNDERSTAND**
5. The project didn't go as we had expected because of some __________ reasons. **SEE**

Translation

VII. Translate the following sentences into English with the words and phrases in brackets.

1. 人工智能是指一种能够像人一样思考的电脑或者由电脑控制的机器。(as…as)

____

____

2. 人工智能研发需要整合诸多学科，如计算机科学、生物学、心理学、语言学、数学、工程学等。(discipline; involve in)

____

____

3. 一些智能系统能够听懂人们与它们交谈时使用的语言。(be capable of; comprehend)

____

____

History of Artificial Intelligence

1 Years ago, artificial beings that could think intelligently appeared as storytelling devices, mostly in fiction as in Mary Shelley's *Frankenstein* or Karel Čapek's *R.U.R. (Rossum's Universal Robots)*. These characters and their fates raised many of the same issues now discussed in the **ethics** of artificial intelligence.

2 The study of mechanical or "formal" reasoning began with philosophers and mathematicians in **antiquity**. The study of mathematical logic contributed directly to Alan Turing's theory of computation, which indicated that a machine, by **shuffling** symbols as simple as "0" and "1", could **simulate** any **conceivable** act of mathematical **deduction**. This insight, that digital computers can simulate any process of formal reasoning, is known as the Church-Turing thesis. This theory, along with concurrent discoveries in **neurobiology**, information theory, and **cybernetics**, aroused the intellectual curiosity of researchers about the possibility of building an electronic brain. Turing proposed that if a human could not distinguish between responses from a machine and a human, the machine could be considered "intelligent". The first work that is now generally recognized as AI was McCulloch and Pitts' 1943 formal design for Turing-complete "artificial neurons".

3 AI research was first started at a workshop at Dartmouth College in 1956. Attendees Allen Newell, Herbert Simon, John McCarthy, Marvin Minsky, and Arthur Samuel became the founders and pioneers of AI research. They and their students produced programs that the press described as "astonishing": Computers were learning checkers strategies (c. 1954) (and by 1959 were reportedly playing better than the average human), solving word problems in **algebra**, proving logical **theorems**, and speaking English. By the middle of the 1960s, the Department of Defense of U.S.A provided adequate funding for the research of AI and overnight laboratories had been set up around the world. AI's founders were positive about the future: Herbert Simon predicted, "Machines

will be capable, within 20 years, of doing any work a man can do." Marvin Minsky agreed, writing, "Within a generation...the problem of creating 'artificial intelligence' will **substantially** be solved."

4 They failed to recognize the difficulty of some of the remaining tasks. Progress slowed and in 1974, in response to the criticism of Sir James Lighthill and ongoing pressure from the U.S. Congress to fund more productive projects, both the U.S. and British governments cut off exploratory research in AI. An "AI winter" came in the next few years, which was a period when it was extremely difficult to obtain funding for AI projects.

5 In the early 1980s, AI research was **revived** by the commercial success of expert systems, a form of AI program that simulated the knowledge and **analytical** skills of human experts. By 1985, the market for AI had arrived at over a billion dollars. At the same time, Japan's Fifth Generation Computer Systems project inspired the U.S and British governments to restore funding for academic research. However, beginning with the **collapse** of the Lisp Machine market in 1987, artificial intelligence once again fell into **disrepute**, and a second, longer-lasting **hiatus** began.

6 The late 1990s and early 21st century witnessed the new application of artificial intelligence in **logistics**, data mining, medical diagnosis, and other areas. This was realized because of increasing computational power, greater emphasis on solving specific problems, new ties between AI and other fields (such as statistics, economics, and mathematics), and a **commitment** by researchers to mathematical methods and scientific standards. One example is that Garry Kasparov who was a **reigning** world chess champion was defeated by Deep Blue, the first computer chess-playing system in 1997.

7 In 2011, the two greatest *Jeopardy!* champions, Brad Rutter and Ken Jennings, were defeated by a significant margin by a question-answering system called Watson, a machine produced by IBM in a *Jeopardy!*'s quiz show exhibition match. Amazing advances in machine learning (ML) and perception were realized due to faster computers, **algorithmic** improvements, and access to large amounts of data; data-hungry deep learning methods started to dominate accuracy **benchmarks** around 2012. The Kinect, which provides a 3D body-motion **interface** for the Xbox 360 and the Xbox One, uses algorithms that emerged from **lengthy** AI research as intelligent personal assistants do in smart phones. In March 2016, AlphaGo won four out of five games of Go in a match with Go champion Lee Sedol, becoming the first computer Go-playing system to beat a professional Go player without **handicaps**. In the 2017 Future of Go Summit, Ke Jie, who at the time continuously held the world No. 1 ranking for two years, lost all his three matches with AlphaGo. This marked the completion of a significant **milestone** in the development of artificial intelligence as Go is an extremely complex game, more so than chess.

8 According to Bloomberg's Jack Clark, 2015 was regarded as a landmark year for artificial intelligence, with an obvious increase in the number of software projects that use AI within Google from a "sporadic usage" in 2012 to more than 2,700 projects. Clark also presents factual

data revealing the fact that since the year of 2011 error rates in image processing tasks have fallen dramatically. He **attributes** this **to** an increase in affordable neural networks, due to a rise in cloud computing **infrastructure**, and an increase in research tools and data sets. Other cited examples include the Skype system developed by Microsoft which can automatically translate from one language to another and Facebook's system that can "picture" images to blind people. In a 2017 survey, one in five companies reported that they had "**incorporated** AI in some offerings or processes".

Notes

Karel Čapek born in Malé Svatoňovice, Austria-Hungary (today Czech Republic), a playwright, novelist, and essayist. He is best known for his science fiction play *R. U. R.* (*Rossum's Universal Robots*).

Alan Turing an English computer scientist, mathematician, logician, philosopher, and theoretical biologist, who was highly influential in the development of theoretical computer science. He is widely considered to be the father of theoretical computer science and artificial intelligence.

McCulloch and Pitts Warren McCulloch (an American neurophysiologist and cybernetician) and Walter Pitts (an American logician) proposed the first artificial neuron in 1943, the model specifically targeted as a computational model of the "nerve net" in the brain.

Allen Newell an American researcher in computer science and cognitive psychology. He contributed to the information processing language and artificial intelligence.

Herbert Simon an American economist and political scientist. He received the Nobel Prize in Economics in 1978 and he was among the pioneers of several modern-day scientific domains such as artificial intelligence, information processing, decision-making, and complex systems.

Marvin Minsky an American cognitive scientist concerned largely with the research of artificial intelligence, who is the co-founder of MIT's AI laboratory.

Arthur Samuel an American pioneer in the field of computer gaming and artificial intelligence.

Sir James Lighthill a British applied mathematician known for his pioneering work in the field of aeroacoustics.

Lisp Machine a general-purpose computer designed to efficiently run Lisp as the main software and programming language, usually via hardware support.

Deep Blue a chess-playing computer developed by IBM. It is known for being the first computer chess-playing system to win both a chess game and a chess match against a reigning world champion under regular time control.

Watson a question-answering computer system capable of answering questions posed in natural language.

Brad Rutter the highest-earning contestant on the U.S. game show *Jeopardy!* and also the highest-earning American game show contestant of all time.

Ken Jennings an American game show contestant and actor. He holds the record for the longest winning streak on *Jeopardy!* and as being the second highest-earning contestant in the game show history.

Lee Sedol a South Korean professional Go player of 9 dan rank, who was defeated by the computer program AlphaGo in a 1–4 series in 2016.

Future of Go Summit held by the Chinese Go Association. It featured five Go games involving AlphaGo and top Chinese Go players, as well as a forum on the future of AI.

Ke Jie a Chinese professional Go player of 9 dan rank. He has held the No. 1 rank in the world since 2014.

Bloomberg Bloomberg News is an international news agency headquartered in New York, the United States. Content produced by Bloomberg News is disseminated through Bloomberg Terminals, Bloomberg Television, Bloomberg Radio, etc.

Jack Clark an American television game show host and announcer.

Google an American multinational technology company that specializes in Internet-related services and products, which include online advertising technologies, search engine, cloud computing, software, and hardware.

Words and Expressions

ethic /'eθɪk/ *n.* (ethics) [pl.] moral principles that control or influence one's behavior 道德准则，伦理标准

antiquity /æn'tɪkwəti/ *n.* ancient times, especially the historic period preceding the Middle Ages in Europe 古代

shuffle /'ʃʌfl/ *v.* to move paper or things into different positions or a different order 打乱次序

simulate /'sɪmjuleɪt/ *v.* to create particular conditions that exist in real life using computers, models, etc., usually for study or training purposes 模拟

conceivable /kən'siːvəbl/ *adj.* capable of being imagined; that you can imagine or believe 可想象的；可信的

deduction /dɪ'dʌkʃn/ *n.* the process of using information you have in order to understand a particular situation or to find the answer to a problem 演绎；推理，推论

neurobiology /ˌnjʊərəʊˌbaɪ'ɒlədʒi/ *n.* the study of the anatomy, physiology, and biochemistry of the nervous system 神经生物学

cybernetics /ˌsaɪbə'netɪks/ *n.* the scientific study of communication and control, especially concerned with comparing human and animal brains with machines and electronic devices 控制论；神经机械学

algebra /'ældʒɪbrə/ *n.* a type of mathematics in which letters and symbols are used to represent quantities 代数

theorem /'θɪərəm/ *n.* a rule or principle, especially in mathematics, that can be proved to be true （尤指数学）定理

substantially /səb'stænʃəli/ *adv.* mainly; in most details, even if not completely 基本上；大体上；总的来说

revive /rɪ'vaɪv/ *v.* to make something start being used or done again 重新使用；重新做

analytical /ˌænə'lɪtɪkl/ *adj.* using a logical method of thinking about something in order to understand it 分析的，解析的

collapse /kə'læps/ *n.* a sudden failure of something, such as an institution, a business, or a course of action 突然失败；崩溃

disrepute /ˌdɪsrɪ'pjuːt/ *n.* the fact that somebody/something loses the respect of other people 丧失名誉，坏名声

hiatus /haɪ'eɪtəs/ *n.* a break in activity when nothing happens 间断，停滞

logistics /lə'dʒɪstɪks/ *n.* the business of transporting and transfering goods 物流

commitment /kə'mɪtmənt/ *n.* the desire to work hard and give your energy and time to a job or an activity 奉献，投入

reigning /'reɪnɪŋ/ *adj.* being the most recent winner of a competition 现任的；本届的

algorithmic /ˌælgə'rɪðmɪk/ *adj.* of or relating to or having the characteristics of an algorithm 算法的；规则系统的

benchmark /'bentʃmaːk/ *n.* something which can be measured and used as a standard that other things can be compared with 基准，参照；标准检测程序

interface /ˈɪntəfeɪs/ *n.* the way a computer program presents information to a user or receives information from a user, in particular the layout of the screen and the menus (人机)界面

lengthy /ˈleŋkθi/ *adj.* very long, and often too long, in time or size 很长的，冗长的

handicap /ˈhændikæp/ *n.* an event or situation that laces you at a disadvantage and makes it harder for you to do something 不利条件，障碍

milestone /ˈmaɪlstəʊn/ *n.* an important event in the history or development of something or someone 里程碑；重大事件；重要阶段

attribute…to to say that something is because of somebody/something 归因于

infrastructure /ˈɪnfrəstrʌktʃə(r)/ *n.* the basic system and services that are necessary for a country or an organization to run smoothly 基础设施，基础建设

incorporate /ɪnˈkɔːpəreɪt/ *v.* to include something so that it forms a part of something 包含，使并入

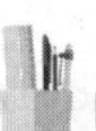

Useful Terms

artificial neuron 人工神经

data mining 数据挖掘

data-hungry deep learning method 数据饥饿深度学习法

sporadic usage 零星使用

cloud computing 云计算

Exercises

Comprehension Check

I. Identify the paragraph from which the information is derived.

1. ______________ It was predicted by Herbert Simon that machines would be able to do what a man could do within 20 years.
2. ______________ The success of expert systems inspired the restoration of AI research.
3. ______________ A machine could be regarded as "intelligent" if human beings couldn't distinguish its responses from a human's.
4. ______________ The application of AI in communication can help describe images to blind people.
5. ______________ Artificial beings first appeared in fiction, which later led to hot issues in the ethics of artificial intelligence.
6. ______________ The success of AlphaGo against the top Go players suggested a tremendous breakthrough of AI because Go is such a complex game.
7. ______________ AI was widely used in many areas because of its great computational power and emphasis on problem-solving ability.
8. ______________ The governments reduced investment in the scientific research of AI.
9. ______________ Jack Clark presents factual data to reveal the significant decline in error rates in image processing tasks since 2011.
10. ______________ McCulloch and Pitts' 1943 formal design for Turing-complete "artificial neurons" was the beginning of AI.

II. Fill in the table below to summarize the history of artificial intelligence according to the text. An example has been given.

No.	Time	Events
e.g.	*In 1956*	*The research of AI was born at a workshop at Dartmouth College.*
1.	In the middle of 1960s	
2.	In 1974	
3.	In the early 1980s	
4.	By 1985	

(Continued)

No.	Time	Events
5.	In 1987	
6.	In the late 1990s and early 21st century	
7.	In 1997	
8.	In 2011	
9.	In 2012	
10.	In 2015	
11.	In 2016	
12.	In 2017	

Vocabulary Building

III. Fill in the following blanks with the words and phrases given in the box. Change the form if necessary.

deduction	strategy	issue	distinguish	simulate
collapse	revive	be known as	due to	attribute...to

1. The process models and the ______________ results can be of some help for the experts in the future research.
2. The ethic value has always been a hot ______________ in the research of cloning.
3. Singapore ______________ "Garden City" because of its clean and beautiful streets and gardens.
4. We need several new marketing ______________ to promote the sales of our new products.
5. In an interview last week, Henry ______________ his academic achievement ______________ his motivation and determination.
6. It is said that science majors are better in logical ______________ than students of liberal arts.
7. To the delight of the engineers, the ______________ of the old systems opened up new possibilities.
8. Over the past five years, the software company ______________ itself as a leading body in the development of software systems.

9. ______________ the maloperation of the machine, there have been serious consequences.

10. There is no doubt that grades will be improved if interest in learning can ______________.

Text C

Current State of Artificial Intelligence

1 Nowadays, AI is playing an **indispensable** role in whatever we do: It is used in modern video games to control the moves of NPCs (non-player characters); it helps learners achieve better learning results; it provides accurate **statistical** information for the police to determine the identity of the criminal. Much of modern technology has, somewhere in the background, some form of AI or machine learning at work, making predictions and decisions, and turning inputs into outputs. **Ubiquity** has made AI somewhat unnoticeable, and AI systems have been hidden, as the result of cloud computing and connected devices, on the edges of our computing efforts.

adj. 不可或缺的
adj. 统计的，统计学的
n. 到处存在，普遍存在

2 Two different models of using AI and machine learning can help you better understand what I mean. Machine learning is being used in the most popular smart phone operating systems in the world, but the methods and **architectures** in these two operating systems are different.

n. 体系结构

3 Android, the operating system used by most smart phones, is written by Google. **Leveraging** the strengths of its maker, Android's use of AI includes using the device as a sort of **appendage**, a sensor package that records, measures, and collects data, which is then sent upstream to servers that use billions of data points collected from millions of users as input for machine learning systems. These collected data sets are then used to produce **weights** for the machine learning system that analyzes photos and tries to understand what the photos suggest. Your photos are both included in Android's larger data set and analyzed against your other photos. When you ask an

v. 最优化使用
n. 附加物，附属物
n. 权重；重量

Android phone to sort out you pictures containing "food", the system is functioning rapidly in the background on an extensive set of complex data exchanges between your local phone and Google's servers, comparing your photos to the billions in its "photos" data set via its machine learning system, and finally, you can view in your phone the pictures that the AI decided were most likely to be related to the concept of "food".

4 This **methodology** has several strong points and weak points. Since Google has to compare billions of photos, and millions of engineers are working hard to help it train its AI, the decisions that the AI makes are generally acceptable. You can do complicated **queries**, such as "Show me photos from the North with a snowman," and the AI will likely succeed in doing just that. Because the system is always **iterating** on itself, learning new weights as new photos are stored and described by people, new objects and events are added to the recognition engine as well. On the other hand, because it is using "public" training sets, and building its predictions and decisions in terms of the actions of everyone using the system, bias and prejudice will be introduced to the system to the same degree as it is present in public. Among the several examples of this surfacing, the most horrifying was the Google Assistant labeling photos of black people as "gorillas".

n. 方法；原则

n. 询问，疑问

v. 迭代，重复操作

5 In contrast, Apple has chosen to build a different model of its AI and machine learning efforts. The analysis and weighting of your photos (as well as other data, but Photos is the easiest category to explain) happen locally, on the devices themselves. If you have an Apple product, you can do similar searches as on an Android phone, for example, "Show me pictures of food." But instead of the weighting and training of the machine learning system happening on Apple's servers somewhere, it all happens locally on the devices. Apple installs models and weights from training sets that it has worked on **remotely** to your phone, but your data and pictures aren't included in that data set. Your local devices use the same machine learning algorithms to contain your photos in Apple's preset weights, but those aren't then sent to Apple's servers to influence others' analysis.

adv. 远程地，远距离地

6 This is also far from perfection, although it is using a different one from Google's approach. Because each data set is analyzed locally,

decision-making cannot be shared as it is with Google. This means that each device has to do the computing heavy lifting itself, rather than relying on remote servers for heavy workload. If you've ever **reinstalled** iOS and are confused why your battery life is terrible on the first day or so and Settings reports that Photos is using more battery than everything else combined, this would explain why. If the system doesn't have a **preexisting** set of search indexes, it will create one via the AI by burning battery life. It also means that each device might have slightly different indexing instead of identical libraries across devices since it's happening entirely locally to the individual machine.

reinstalled *v.* 重新安装

preexisting *adj.* 先前存在的；预设的

7 The advantages of localized machine learning can be seen in enormous gains in privacy and security of information. Statistics show that the surface attacks for the data and risk of privacy issues are hugely reduced if there is no sending any photo and data from servers to clients and if providers don't need to store and host data. Let's take the example of photo library. Apple doesn't have access to the photos directly because of the methodology it uses to store and transmit data from your phone to its iCloud servers. According to the iOS 12 security paper, for instance, "Each file is broken into **chunks** and **encrypted** by iCloud using AES-128 and a key derived from each chunk's contents utilizes SHA-256. The keys and the file's **metadata** are stored by Apple in the user's iCloud account. The encrypted chunks of the file are stored, without any user-identifying information or the keys, using both Apple and third-party storage services."

chunks *n.* 大块；厚块

encrypted *v.* 加密

metadata *n.* 元数据

8 This ensures that Apple doesn't have information that may place a user's privacy in danger, even though it might be not good enough for certain machine learning tasks. It is evident why this methodology difference might be of interest to libraries. As libraries and library vendors make efforts to develop AI and machine learning systems, we should be very alert to the privacy implications of collecting and storing data needed to train and update those systems. Along with existing systems where we **outsource** data collection and retention to vendors, libraries need to be very conscious of the mechanisms by which that data is protected and how it may be shared with others through training sets. Therefore, it is necessary for libraries to provide local analysis in the style of Apple and iOS.

outsource *v.* 交外办理（外包）

9 From what has been discussed, we can draw a clear picture that two different methodologies are employed to do work using AI systems and focus on object and image recognition in photos as the function of the machine learning system. There have been **a multitude of** uses to which AI and machine learning systems are being applied in modern technology. Very broadly, the uses of AI could be categorized as "analysis and **synthesis** of media" in current tech, as so many systems are being designed to do recognition and **semantic** analysis work. Android and iOS mentioned above are good examples of common use of AI and machine learning in analyzing photos for objects. AI systems can be trained to recognize locations from photos we have taken, not only in terms of the subject of the photo, but also in terms of the location of the photographer (based on angle, geography, and more). These systems could be extremely useful in making the processing of **archives** and collections more quickly findable.

a multitude of 大量的，海量的
synthesis *n.* 综合；结合
semantic *adj.* 语义的，语义学的
archives *n.* 档案；归档

10 Similar types of systems are being developed for video, where a series of photographs that constitute video are analyzed and processed for a variety of different pieces of information, depending on the purpose of creating the video. These can be helpful, in the case of something like HomeCourt, an iOS app that watches video of players on a basketball court and tracks position, form, shooting percentages, and more in order to help players learn from their performances. Or they can be potentially harmful when they are used to enable nearly real-time tracking of individuals through a store, mall, or even down city streets.

Exercises

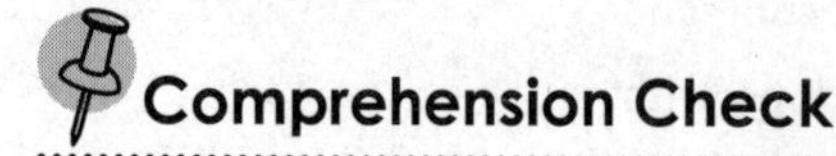

Comprehension Check

I. Answer the following questions according to the text.

1. How important is artificial intelligence in modern video games?
2. What are the two most popular smart phone operating systems in the world currently?
3. What will happen if you ask an Android phone to show you pictures of beaches?

4. What is one of the disadvantages of iPhones?

5. How can the attack surface for the data and risk of privacy issues be reduced?

II. Read the following statements carefully and decide whether they are true (T) or false (F) without turning back to check the text.

1. ______________ People may neglect the role of AI just because of its extensive use in daily life.

2. ______________ A majority of smart phones are using the iOS operating system.

3. ______________ Google has millions of people working on how to train the Android system so the AI system can make good decisions.

4. ______________ Google Assistant labels photos of black people as "gorillas" because the system fails to detect the difference between a human and a real gorilla.

5. ______________ The battery on an Android smart phone is generally better than on an iPhone.

III. Discuss the following questions based on your understanding of artificial intelligence.

1. Why does the development of AI in China thrive?

2. It is said that nowadays, artificial intelligence is doing less than we can think; however, it will surely do more than we can imagine. Can you explain how a machine can learn to drive automatically?

3. The rapid development of AI will have huge impact on labor divisions and job market. In what fields will AI influence employment and organization in the industrial working environment of the future?

Writing Skills

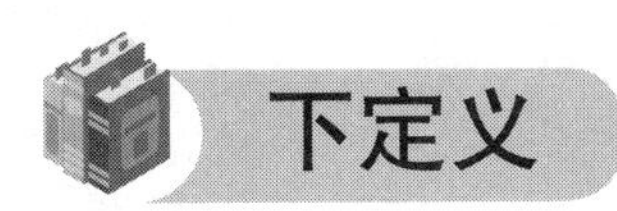

下定义

下定义（defining）是对某种事物的特征、某个概念的内涵以及某种物体的属性进行简单的说明，即解释某事物是什么，有什么特征。下定义时，通常把概念或要下定义的事物置于一个大的类别中，并解释该事物与同一类别的其他事物有何不同。下定义法多用于展开段落，可以为读者提供一个直观的概念或者清晰的思路，并且有利于推动后文的发展。

英语中给某一事物、物体或概念下定义的方式包括以下四种：

1. 定语从句（包括限定性和非限定性定语从句）

例 1：Go is a board game which involves abstract strategy between two players and the aim is to surround more territory than the opponent.

2. 非谓语动词（不定式、动名词、分词形式）

例 2：Cricket is a bat-and-ball game played between two teams of 11 players on a field at the center of which is a 20-meter (22-yard) pitch with a wicket at each end, each comprising two bails balanced on three stumps.

3. 句型 define something as... 或者 something is defined as...

例 3：Sociology can be defined as one branch of science, which studies the development and principles of social organization.

4. 描述功能

例 4：Smoking signals, carrier pigeons, and riders on horseback used to be the major means of sending important messages. But today, what we need to do is just to pick up our mobile phones.

Define the following technical terms.

1. big data

____

2. cloud computing

____

3. driverless vehicle or unmanned vehicle

____

Translating Skills

被动语态

汉语和英语都有被动语态（passive voice）。相对于英语中的被动语态而言，汉语的被动语态会根据句子内涵、强调的主题内容等产生多种变化形式，主要原则是正面、肯定的信息尽量采用主动的表达形式，而负面、否定的信息则会采用被动的表达形式。除了使用“被”字表示之外，汉语还会使用“受（到）”“遭（到）”“获”“挨”等辅助词语来表示被动。英语中只有直接、简单的被动语态，即 be done (by) 或者分词短语 done (by) 形式。

值得一提的是，汉语句子可以没有主语，但是英语句子必须要有主语。因此，在汉译英的过程中，当遇到以下三种情况时，需要使用被动语态进行翻译：

1. 原文中没有主语（不知道动作发出者是谁、没有必要提及动作发出者）

例 1：机器学习在世界上最流行的智能手机操作系统中得到了广泛应用，但是这两种操作系统的方法和体系结构是不同的。

Machine learning is being used in the most popular smart phone operating systems in the world, but the methods and architectures in these two operating systems are different.

2. 原文中出现了表示被动语态的词语，如“被”“受（到）”“遭（到）”“获”“挨”等

例 2：计算机的一个分支被称为人工智能，即力求创造出同人一样智能的电脑或机器。

A branch of computer science named artificial intelligence pursues creating the computers or machines as intelligent as human beings.

3. 强调动作的承受者

例 3：到 20 世纪 60 年代中期，美国国防部大力资助了国内的研究，并在世界各地建立了多个实验室。

By the middle of the 1960s, research in the U.S. had been heavily funded by the Department of Defense and laboratories had been established around the world.

Translation at Sentence Level

I. Translate the following sentences into Chinese.

1. AI was formally initiated in 1956, when the name was coined, although at that point work had been underway for about five years.

____

____

2. 2015 was regarded as a landmark year for artificial intelligence, with an obvious increase in the number of software projects that use AI within Google from a "sporadic usage" in 2012 to more than 2,700 projects.

____

____

3. All texts written on paper by a pen or on screen by a stylus are readable via the handwriting recognition system, which can "read" the shapes of the letters and change them into editable text.

____

____

4. Ubiquity has made AI somewhat unnoticeable, and AI systems have been hidden, as the result of cloud computing and connected devices, on the edges of our computing efforts.

____

____

5. Along with existing systems where we outsource data collection and retention to vendors, libraries need to be very conscious of the mechanisms by which that data is protected and how it may be shared with others through training sets.

____

____

II. Translate the following sentences into English.

1. 目前在美国，人工智能技术由至少 16 个独立机构管理。

____

____

2. 人工智能将会被更广泛、更深入地融入工业生产过程和消费产品中。

____

____

3. 在过去的 15 年里，人工智能取得的显著进展已经对北美的许多大城市产生了影响。

____

____

4. 据说，大批劳动力将会被人工智能和机器学习所取代。

____

____

5. 自从第一台计算机发明以来，智能计算机的前景就一直吸引着人们。

____

____

Translation at Paragraph Level

English to Chinese Translation

①Controlling vehicles is brought down to earth by Ernest D. Dickmanns in *Vehicles Capable of Dynamic Vision: A New Breed of Technical Beings?*. ②He surveys two decades of developments in road vehicle guidance. ③These advances have been fueled by new instrumentation, four orders of magnitude of increase in computing power, and combination of AI methods with some from system dynamics and control engineering. ④This research has resulted in systems that can control vehicles on public freeways at speeds beyond 130 km/h. ⑤Automobile traffic gives rise to the driving example used by Timothy Huang and Stuart Russell in their paper "Object Identification: A Bayesian Analysis with Application to Traffic Surveillance". ⑥They define the notion of an appearance probability that specifies expectations of how an object might appear later, given its current appearance. ⑦This is used as the basis for a Bayesian analysis which, when applied to the problem of traffic analysis, provides both good explanatory power and high levels of performance.

本段一共七句话。段落英译汉和段落汉译英是两个方向相反的翻译处理过程，但原理和方法却是基本一致的，都需要考虑英语、汉语在段落表达层面的特点，都需要进行句子层面的结构分析，都需要确保译文的意思与原文一致，都需要符合目的语的语法要求和表达习惯，使译文通顺流畅。

第一句中含有被动语态，原主动发出者（逻辑主语）为 Ernest D. Dickmanns，谓语动词为 bring down to earth，逻辑宾语为 controlling vehicles，整句话的逻辑思路是“Ernest D. Dickmanns brings sth. down to earth (in his book).”，在翻译成汉语时可以将被动语态译为“某人提出什么”。全句可以译为：“Ernest D. Dickmanns 在《具有动态视觉能力的车辆：一种新的技术生物？》一书中明确提出了控制车辆这一概念。”第二句是一个简单句，句子主干分别是主语 he、谓语动词 survey 和宾语 two decades of developments。全句可以译成：“他调查了 20 年来道路车辆导航的发展。”

第三句也是一个含有被动语态的句子，逻辑主语是 by 后面的三个并列名词词组，谓语动词是 fuel，逻辑宾语为 these advances。由于主语过长，宾语相对而言较短，为了避免头重脚轻，可以把宾语前置变成主语，句子的语态改为被动语态，原来的谓语动词改为 be fueled by，因此这句话既可以还原成主动语态，译成：“新的仪器设备、计算能力四个数量级的提高以及人工智能方法与系统动力学和控制工程的一些方法相结合，推动了这些进展”，也可以保留原文的被动语态，译成：“这些发展要得力于新的仪器设备、计算能力四个数量级的提高以及人工智能方法与系统动力学和控制工程的一些方法的结合。”第四句是一个复合句，内含一个定语从句，句子主干分别为主语 this research、谓语动词 result in 和宾语 systems，后跟一个定语从句修饰 systems。全句可以译成：“这项研究已经产生了一种可以控制高速公路上时速超过 130 公里的车辆的系统。”

第五句中的被动语态是一个过去分词短语（used by），修饰前面的名词 driving example，可以直接翻译成：“Timothy Huang 和 Stuart Russell 在他们的论文《对象识别：贝叶斯分析及其在交通监控中的应用》中使用的驾驶实例正是这一佐证”。第六句也是一个复合句，带有一个由 that 引导的同位语从句，全句可译成：“他们定义了外观概率的概念，该概念说明了一个对象在给定当前外观的情况下，以后可能会如何出现的期望。”最后一句中的被动语态与前几句不同。由于缺少逻辑主语，即 use 的主动发出者未被提及，本句可以采用被动语态，译为：“这被用作贝叶斯分析的基础，当应用于流量分析问题时，它既提供了良好的解释力，又提供了高水平的性能。”

III. Translate the following paragraph into Chinese.

While AI and machine learning systems will provide untold benefits for libraries, the risks and concerns that have arisen over the last several years in regard to AI systems should give us significant pause. AI is only as good as its training data and the weighting that is given to the system as it learns to make decisions. If that data is biased, contains bad examples of decision-making, or is simply collected in such a way that it isn't representative of the entirety of the problem set that will be asked of the system in the end, that system is going to produce broken, biased, and bad outputs. These may reflect social issues, where data could cause the

AI system to be racist in its decision-making, or classist, or sexist…any sort of non-balanced inputs can cause the outputs to reinforce the negative. We've seen this from the largest technology companies in the world, and unless we are very careful about how AI can be implemented in library work, we risk doing serious damage to serve our patrons.

____

____

____

____

____

____

Chinese to English Translation

①随着全球新一轮科技革命和产业变革的不断深入推进，5G 已成为世界数字经济发展战略中的优先发展领域。②将 5G 视为推进供给侧结构性改革的新动能、振兴实体经济的新机遇、建设制造强国和网络强国的新引擎，大力推进 5G 建设已经提升至国家战略高度。③5G 融合大数据 / 人工智能等通用技术，将全面构建我国数字经济的关键基础设施，赋能智慧交通、智慧城市、智慧健康医疗、智能制造、超高清视频等各行各业，推动社会生活方式与生产方式的深刻变革。

本段一共三句话。鉴于中英文在段落表达方面的差异，翻译前需要对句子的逻辑关系进行分析，翻译时需要先分析句子结构，采用合适的方法确定英文句的主谓结构，再处理细节内容的转换，翻译完成后需要检查英文句子的主语、谓语在人称、数量上是否一致，尤其需要注意检查谓语动词的语态、时态和语气是否正确。

第一句的主语是后半句中的“5G”，谓语是系动词“成为”，后接表语“优先发展领域”；前半句为状语，可以采用介词短语“with + 名词词组”或者用 as 引导的状语从句进行翻译。因此，本句可以处理为“状语 + 主系表”结构：“With the new round of global scientific and technological revolution and the continuous deepening of industrial reform / As the new round of global scientific and technological revolution comes and industrial reform deepens continuously, 5G has become a priority in the development strategy of the world's digital economy. ”。

第二句的主干部分为“将 5G 视为……”，由于此句为无主句，即没有说明“谁”将 5G 视为，因此在译成英语时需要调整语态，将“5G”改为主语，谓语动词“视为”改为被动语态。第二句可以译为：“5G will be regarded as a new driving force for promoting supply side structural reform, a new opportunity for revitalizing the real economy, and a new engine for building a manufacturing power and a network power.”。

第三句的结构比较复杂，首先主语是一个词组“5G 融合大数据 / 人工智能等通用技术”，谓语动词是两个并列动词“构建”“赋能”，后面接了一个目的状语“推动变革”。由于目的状语与前面两个并列谓语动词在形式和结构上容易混淆，可以考虑单独翻译目的状语，将本来的“动词 + 宾语”结构改为被动语态。另外，“构建”和“赋能”后接较多宾语，并列结构较为复杂，因此可以将“赋能”改用分词作伴随状语，避免使用多个性质不同的 and。因此，本句可以译为：“The integration of 5G and big data, artificial intelligence, and other general technologies, will comprehensively build the key infrastructure of China’s digital economy, enabling intelligent transportation, intelligent city, intelligent healthcare, intelligent manufacturing, ultra-high definition video, and other industries. Therefore, the profound change of social lifestyle and production mode will be promoted.”。

IV. Translate the following paragraph into English.

2019 年 11 月 20 日至 23 日，世界 5G 大会在北京成功举办。本次大会以“5G 改变世界，5G 创造未来”为主题，以“国际化、高端化、专业化”为特色，集聚全球信息通信领域最具影响力的科学家和企业家，以及相关政府的领导人，围绕 5G 领域的技术前沿、产业趋势、创新应用等发表演讲和进行高端对话，为智慧城市、智慧交通与人工智能、大数据、云计算、物联网等网络信息技术的融合发展，提供了发展思路和创新路径，为各界搭建了世界顶尖的 5G 交流合作平台。

____

____

____

____

____

____

Workshop

I. **Choose one computer system or IT product and do research on how it is developed. Then write an outline of your research results and make a five-minute oral presentation to the class.**

Name: ____________________

Country: ____________________

Start Date: ____________________

Founders: ____________________

Main Features: ____________________

Number in Use: ____________________

II. **Fill in the table below to make a comparison between Deep Blue and AlphaGo and discuss with your partners about which one is better.**

Details	Deep Blue	AlphaGo
Founding Time		
Location		
Researchers		
Company		
Fields		
Achievements		
Main Features		

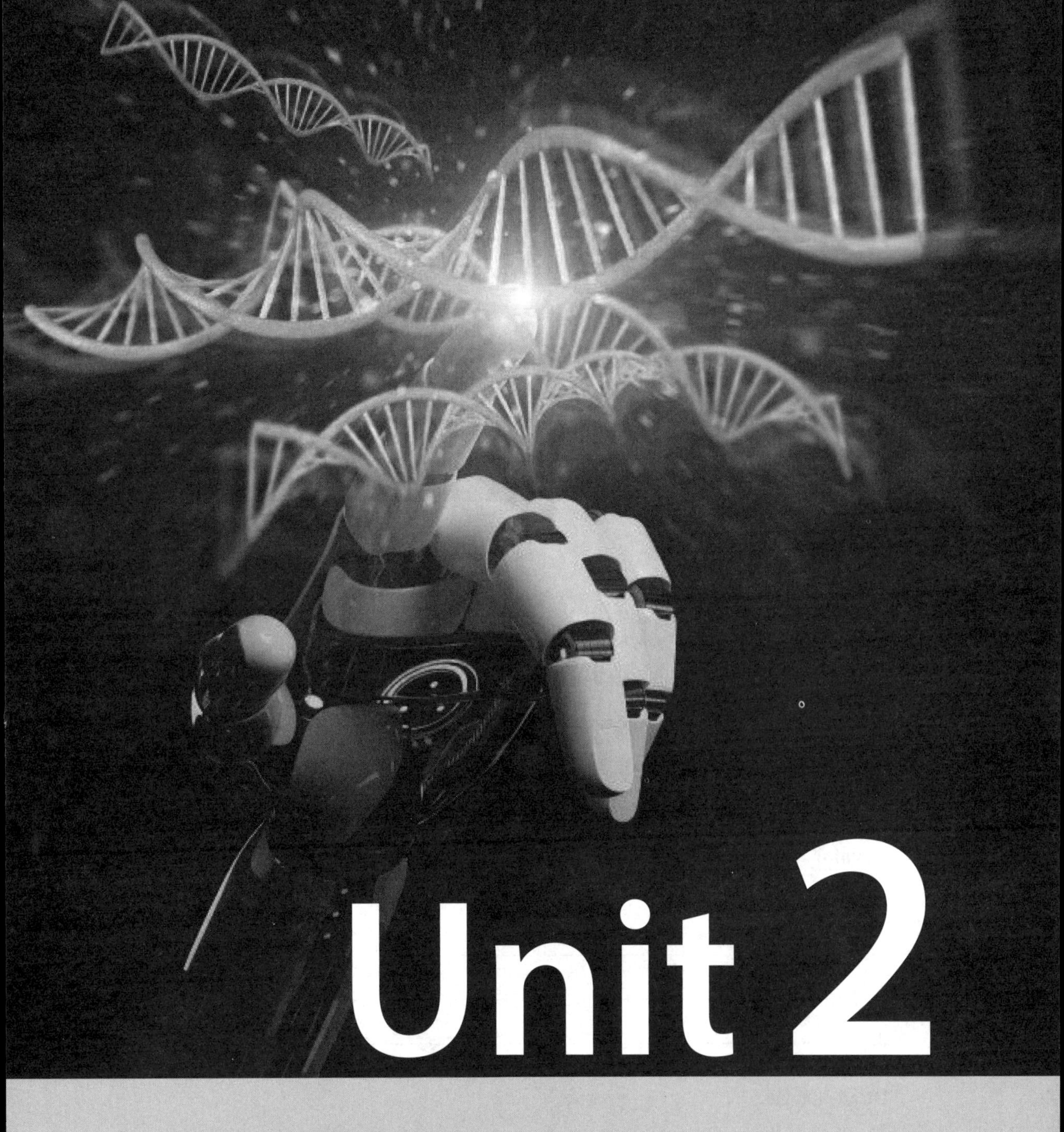

Unit 2

Artificial Intelligence and Machine Learning

Learning Objectives

In this unit, you will learn:

- a brief introduction to machine learning;
- applications and limitations of machine learning;
- basics of machine learning;
- writing skills—the usage of appositives;
- translating skills—conversion of part of speech.

Lead-in

I. Discuss the following questions with your partners.

1. What is the relationship between artificial intelligence and machine learning?
2. How does machine learning work?
3. What are the applications of machine learning?

II. Work in pairs and discuss the advantages and disadvantages of present-day machine learning.

Advantages:

1. ______________________________
2. ______________________________
3. ______________________________

Disadvantages:

1. ______________________________
2. ______________________________
3. ______________________________

Text A

An Introduction to Machine Learning

1 In 1959, Arthur Samuel, an American pioneer in the field of computer gaming and artificial intelligence, coined the term "machine learning". Machine learning is an application of artificial intelligence that provides systems with the ability to automatically learn and improve from experience without being explicitly programmed. Machine learning focuses on the development of computer programs that can **access** data and use it for learning by themselves. The process of learning begins with observations or data, such as examples, direct experience, or instruction, in order to look for patterns in data and make better decisions in the future based on the examples that we provide. The primary aim is to allow the computers to learn automatically without human **intervention** or assistance and **adjust** actions accordingly.

Algorithms

2 Machine learning algorithms are often categorized as supervised, unsupervised, and reinforcement.

3 Supervised machine learning algorithms can apply what has been learned in the past to new data using labeled examples to predict future events. Starting from the analysis of a known training data set, the learning algorithm produces an inferred function to make predictions about the **output** values. The system is able to provide targets with any new input after sufficient training. The learning algorithm can also compare its output with the correct, intended output and find errors in order to modify the model accordingly.

4 In contrast, unsupervised machine learning algorithms are used when the information used to train is neither classified nor labeled. Unsupervised learning studies how systems can infer a function to describe a hidden structure from unlabeled data. The system doesn't figure out the right output, but it explores the data and can draw inferences from data sets to describe hidden structures from unlabeled data.

5 Reinforcement machine learning algorithms are a learning method that interacts with its environment by producing actions and discovers errors or rewards. Trial-and-error search and delayed reward are the most relevant characteristics of reinforcement learning. This method allows machines and software agents to automatically determine the ideal behavior within a specific

context in order to maximize its performance. Simple reward feedback is required for the agent to learn which action is best; this is known as the reinforcement signal.

Applications

6 Machine learning enables analysis of massive quantities of data. While it generally delivers faster, more accurate results in order to identify profitable opportunities or dangerous risks, it may also require additional time and resources to train it properly. Combining machine learning with AI and cognitive technologies can make it even more effective in processing large volumes of information.

7 In 2006, the online movie company Netflix held the first "Netflix Prize" competition to find a program to better predict user preferences and improve the accuracy on its existing CineMatch movie recommendation algorithm by at least 10%. A joint team made up of researchers from AT&T Labs Research in **collaboration** with the teams Big Chaos and Pragmatic Theory built an ensemble model to win the **grand prize** in 2009 for $1 million. Shortly after the prize was awarded, Netflix realized that viewers' ratings were not the best **indicators** of their viewing patterns ("everything is a recommendation") and it changed its recommendation engine accordingly. In 2010, *The Wall Street Journal* wrote about the firm Rebellion Research and its use of machine learning to predict the financial crisis. In 2012, co-founder of Sun Microsystems, Vinod Khosla, predicted that 80% of medical doctors' jobs would be lost in the next two decades to automated machine learning medical diagnostic software. In 2014, it was reported that a machine learning algorithm had been applied in the field of art history to study fine art paintings, and that it may have revealed previously unrecognized influences between artists.

Limitations

8 Although machine learning has been **transformative** in some fields, machine learning programs often fail to deliver expected results. Reasons for this are numerous: lack of (suitable) data, lack of access to the data, data bias, privacy problems, badly chosen tasks and algorithms, wrong tools and people, lack of resources, and evaluation problems. In 2018, a self-driving car from Uber failed to detect a pedestrian, who was killed after a collision. Attempts to use machine learning in healthcare with the IBM Watson system failed to deliver results even after years of time and billions of investments.

9 Machine learning approaches in particular can suffer from different data biases. A machine learning system trained on current customers only may not be able to predict the needs of new customer groups that are not **represented** in the training data. When trained on man-made data, machine learning is likely to **pick up** the same **constitutional** and unconscious biases already present in society. Language models learned from data have been shown to contain human-like biases. Machine learning systems used for criminal risk **assessment** have been found to be biased against black people. In 2015, Google Photos would often **tag** black people as **gorillas**, and in

2018, this still was not well resolved, but Google reportedly was still using the **workaround** to remove all gorillas from the training data, and thus it was not able to recognize real gorillas at all. Similar issues with recognizing non-white people have been found in many other systems. In 2016, Microsoft tested a chatbot that learned from Twitter, and it quickly picked up **racist** and **sexist** language. Because of such challenges, the effective use of machine learning may take longer to be adopted in other domains. Concerns for reducing bias in machine learning and **propelling** its use for human good are increasingly expressed by artificial intelligence scientists, including Fei-Fei Li, who reminds engineers: "There's nothing artificial about AI...It's inspired by people, it's created by people, and most importantly, it impacts people. It is a powerful tool we are only just beginning to understand, and that is a profound responsibility."

Notes

"Netflix Prize" competition an open competition for the best collaborative filtering algorithm to predict user ratings for films, based on previous ratings without any other information about the users or films, i.e. without the users or films being identified except by numbers assigned for the contest. The competition was held by Netflix, an online DVD-rental and video streaming service, and was open to anyone who is neither connected with Netflix (current and former employees, agents, close relatives of Netflix employees, etc.) nor a resident of certain blocked countries (such as Cuba or DPRA).

AT&T Labs the research and development division of AT&T. It employs some 1,800 people in various locations, including: Bedminster, Middletown, Manhattan, Warrenville (Integrated Test Network), Austin, Atlanta, San Francisco, San Ramon, and Redmond (Mobility Test Center). AT&T Labs Research, the 450-person research division of AT&T Labs, is based in the Bedminster, Middletown, San Francisco, and Manhattan locations. AT&T Labs traces its history from AT&T Bell Labs. Its research areas are traditionally associated with networks and systems, ranging from the physics of optical transmission to foundational topics in computing and communications. Other research areas address the technical challenges of large operational networks and the resulting large data sets.

Sun Microsystems an American company that sells computers, computer components, software, and information technology services and creates the Java programming language, the Solaris Operating System (SOS), the Z Film System (ZFS), the Network File System

(NFS), and Scalable Processor Architecture (SPARC). It contributes significantly to the evolution of several key computing technologies, among which are Unix, RISC processors, thin client computing, and virtualized computing. It was founded on February 24, 1982. At its height, the headquarters was in Santa Clara, California (part of Silicon Valley), on the former west campus of the Agnews Developmental Center.

Vinod Khosla

an Indian American billionaire engineer, businessman, and venture capitalist. He is a co-founder of Sun Microsystems and Khosla Ventures. In 2014, *Forbes* named him amongst the 400 richest people in the world.

Uber failed to detect a pedestrian

On March 18, 2018, a Uber self-driving car made a collision which caused the death of a pedestrian. On December 20, 2018, Uber returned self-driving cars to the roads in public testing in Pittsburgh, Pennsylvania. Uber said that it received authorization from the Pennsylvania Department of Transportation, and that it was also pursuing deploying cars on roads in San Francisco, California, and Toronto.

Google Photos

a photo sharing and storage service developed by Google. It was announced in May 2015. Google Photos gives users free, unlimited storage for photos. Users can search for anything in photos, with the service returning results from three major categories: people, places, and things. Google Photos recognizes faces, grouping similar ones together (this feature is only available in certain countries due to privacy laws), geographic landmarks (such as the Eiffel Tower), subject matter, etc.

Fei-Fei Li

a professor in the Computer Science Department at Stanford University, and co-director of Stanford's Human-Centered AI Institute. She served as the director of Stanford's AI Lab from 2013 to 2018. And during her sabbatical from Stanford from January 2017 to September 2018, she was vice president at Google and served as chief scientist of AI/ML at Google Cloud.

Words and Expressions

access /ˈækses/ *v.* to obtain or retrieve information from a storage device, such as a computer 使用，访问（网站、数据等）

intervention /ˌɪntəˈvenʃn/ *n.* the act of getting involved, so as to alter or hinder

			an action, or through force or threat of force 干预
adjust	/ə'dʒʌst/	*v.*	to alter or regulate so as to achieve accuracy or conform to a standard 调整；校准
output	/'aʊtpʊt/	*n.*	the information, results, etc. produced by a computer 输出
collaboration	/kəˌlæbə'reɪʃn/	*n.*	the act of working jointly 合作
grand prize			the top prize given in a contest 大奖
indicator	/'ɪndɪkeɪtə(r)/	*n.*	a measurement or value that gives an idea of what something is like 指示物；标志
transformative	/træns'fɔːmətɪv/	*adj.*	causing or able to cause an important and lasting change in somebody/something 有改革能力的
represent	/ˌreprɪ'zent/	*v.*	to be a member of a group of people and act or speak on their behalf at an event, a meeting, etc. 代表
pick up			to get to know or become aware of, usually accidentally 学会
constitutional	/ˌkɒnstɪ'tjuːʃənl/	*adj.*	connected with the constitution of a country or an organization 宪法的；章程的
assessment	/ə'sesmənt/	*n.*	the act of judging or forming an opinion about somebody/something 评定；核定；判定
tag	/tæg/	*v.*	to fasten a tag onto somebody/something 给……贴上标签
gorilla	/gə'rɪlə/	*n.*	a very large powerful African ape covered with black or brown hair 大猩猩
workaround	/'wəːkəraʊnd/	*n.*	a way in which you can solve or avoid a problem when the most obvious solution is not possible 变通方案
racist	/'reɪsɪst/	*n.*	a person with a prejudiced belief that one race is superior to others 种族主义者
sexist	/'seksɪst/	*n.*	a person who treats others, especially women, unfairly because of their sex, or who makes offensive remarks about them 性别歧视者
propel	/prə'pel/	*v.*	to cause to move forward with force 推进，促进

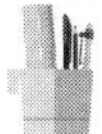

Useful Terms

data set 数据集

CineMatch	电影推荐算法
ensemble model	组合模型
recommendation engine	推荐引擎
supervised machine learning algorithm	监督机器学习算法
unsupervised machine learning algorithm	无监督机器学习算法
reinforcement machine learning algorithm	强化机器学习算法
chatbot	聊天机器人

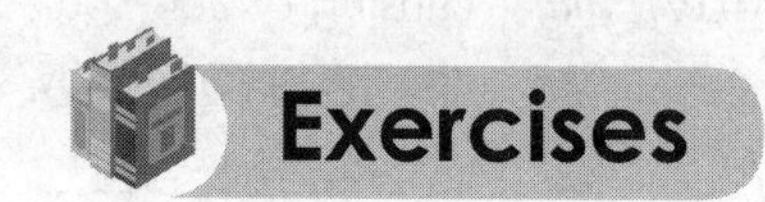

Exercises

Comprehension Check

I. Answer the following questions according to the text.

1. Who coined the term "machine learning"?
2. What is the definition of machine learning?
3. What is the aim of machine learning?
4. How many machine learning algorithms are mentioned? And what are they?
5. What are the applications of machine learning? And how about the limitations?

II. Read the following statements carefully and decide whether they are true (T) or false (F) without turning back to check the text.

1. ______________ Supervised machine learning algorithms can apply what has been learned at present to predict future events.
2. ______________ Semi-supervised machine learning algorithms use both labeled and unlabeled data for training.
3. ______________ Machine learning is combined with AI and cognitive technologies, which can make it even more effective in processing large volumes of information.
4. ______________ Machine learning systems used for criminal risk assessment often recognize black people as criminals.
5. ______________ The effective use of machine learning may take longer to be adopted in other fields.

III. Choose the best answer to each of the following questions according to the text.

1. Machine learning is an application ______________.

   A. that used to be an academic discipline of AI in early days

   B. aiming to guide computers to learn and adjust actions under human assistance

   C. providing AI with the capability to learn by itself and from experience

   D. with observations on data to make a better decision at the moment

2. How can machine learning deal with massive information more effectively?

   A. By using more time and resources to train it properly.

   B. By analyzing massive quantities of data.

   C. By combining itself with AI and cognitive technologies.

   D. By identifying results of profitable opportunities or dangerous risks.

3. Machine learning has been used in some fields EXCEPT ______________.

   A. medical diagnosis

   B. film recommendation

   C. self-driving

   D. financial crisis prediction

4. What may be the reason(s) that machine learning programs fail to deliver expected results?

   A. Lack of appropriate data.

   B. Wrong choices of tasks and algorithms.

   C. Privacy problems.

   D. All of the above.

5. What can we learn from the last paragraph?

   A. A machine learning system can predict the needs of new customer groups that are not represented in the training data.

   B. Data biases may occur when machine learning is trained on man-made data.

   C. Google Photos can now recognize real gorillas.

   D. It takes a short time to use machine learning effectively in every field.

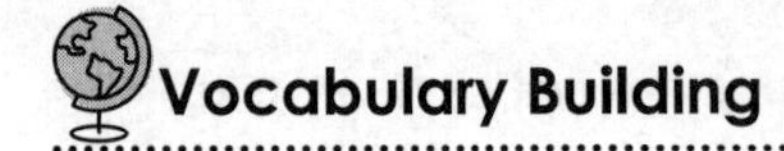

Vocabulary Building

IV. Fill in the following blanks with the words and phrases given in the box. Change the form if necessary.

tag	artificial	propel	categorize	bias
interact with	modify	primary	figure out	pick up

1. One advantage of having ______________ limbs is that you can upgrade them.
2. I don't want to ______________ any bad habits from those people.
3. His misunderstanding of language was the ______________ cause of his other problems.
4. You can ______________ it to suit your needs if you like.
5. The players are all trying to ______________ the audience in a positive way.
6. Our country has been no longer ______________ as a poor and weak nation.
7. Solar sails rely on sunlight to ______________ vehicles through space.
8. No matter what kind of students he encounters, he treats them impartially, without any ______________.
9. Make a list of your child's toys and then ______________ them as sociable or antisocial.
10. If you ______________ a solution to a problem or the reason for something, you succeed in solving it or understanding it.

V. Match the words in the left column with the explanations in the right column.

1. assessment	A. an idea or action intended to deal with a problem or situation
2. approach	B. to recognize somebody/something and be able to say who or what they are
3. represent	C. the act of judging or forming an opinion about somebody/something
4. preference	D. a strong liking
5. identify	E. a member of a group of people, who acts or speaks on their behalf at an event

Word Formation

VI. Fill in the following blanks with the words in capitals. Change the form if necessary. An example has been given.

e.g. *In the absence of prediction and remediation, the best we can do is to be prepared for them.* PREDICT

1. Parents should avoid too much ______________ in their children's education. INTERVENE
2. I hear the students gave the new teacher an unfair ______________. EVALUATE
3. The government's present work has received popular ______________. RECOGNIZE
4. Those around you need your continual guidance, ______________, and encouragement. INSPIRE
5. The dynamic market demands constant changes and ______________. ADJUST

Translation

VII. Translate the following sentences into English with the words and phrases in brackets.

1. 即使尝试了很多次，人们也未能将特定语言与大脑的特定部分联系起来。(particular; fail to)

   ____

   ____

2. 公众越来越关注网络隐私安全。(concern; express)

   ____

   ____

3. 机器学习应用程序与早期的手机开发应用程序不同。(in contrast with)

   ____

   ____

Text B

Applications and Limitations of Machine Learning

1 Some specialists believe that applications of machine learning are, on the one hand, magic boxes capable of doing whatever we want or, **conversely**, are **alien**-like solutions that are useless in everyday life. As it often happens, especially when it comes to new technologies, the truth lies somewhere in the middle.

Applications of Machine Learning

2 One of the most common uses of machine learning is image recognition. There are many situations where you can **classify** the object **as** a digital image. For digital images, the **measurements** describe the outputs of each **pixel** in the image. Machine learning can be used for face detection and **character** recognition.

3 Speech recognition (SR) is the translation of spoken words into text. It is also known as "automatic speech recognition" (ASR), "computer speech recognition" (CSR), or "speech to text" (STT). The applications of speech recognition include voice user interfaces, such as voice dialing, call routing, and domestic appliance control. SR can also be used as simple data entry, preparation of structured documents, speech-to-text processing, and **plane.**

4 Machine learning provides methods, techniques, and tools that can help solving diagnostic and **prognostic** problems in a variety of medical domains. It is being used for the analysis of the importance of clinical **parameters** and of their combinations for prognosis, e.g. prediction of disease progression, for the **extraction** of medical knowledge for outcomes research, for **therapy** planning and support, and for overall patient management. Machine learning is also being used for data analysis, such as detection of **regularity** in the data by **appropriately** dealing with imperfect data, for interpretation of continuous data used in the intensive care unit, and for intelligent alarming resulting in effective and efficient monitoring. The measurements in this application are **typically** the results of certain medical tests (such as blood pressure, temperature, and various blood tests) or medical diagnostics (such as medical images), presence/absence/intensity of various **symptoms**, and basic physical information about the patient (such as age, sex, weight, etc.). On

the basis of the results of these measurements, the doctors **narrow down** the disease **inflicting** the patient.

5 Classification is a process of placing each individual from the population under study in many classes. This is identified as independent variables. Classification helps analysts to use measurements of an object to identify the category to which that object belongs. For example, before a bank decides to **disburse** a loan, it assesses customers on their ability to repay the loan. By considering factors such as customer's earning, age, savings, and financial history, we can do it. This information is taken from the past data of the loan. Hence, Seeker uses it to create a relationship between customer attributes and related risks.

6 Currently, prediction is one of the hottest machine learning algorithms. Let's take an example of **retail**. Earlier we were able to get insights like sales report last month/year/5-years/Christmas. This type of reporting is called historical reporting. But currently, business is more interested in finding out what the sales will be next month/year, etc., so that business can take required decision (related to procurement, stocks, etc.) on time.

7 Information extraction (IE) is another application of machine learning. It is the process of extracting structured information from unstructured data, for example, Web pages, articles, blogs, business reports, and e-mails. The relational database maintains the output produced by the information extraction. Nowadays, IE is becoming a key in big data industry.

Limitations of Machine Learning

8 In business, we can say that machine learning offers a **sophisticated** approach, but there is a limit to the level of improvement possible in analyzing unstructured information. In fact, applications of machine learning:

- Need data or models that have been prepared **manually** by people. And even then the process is not completely automatic. These applications do not learn on their own; someone has to teach them the differences between topics, words, concepts, etc.
- Require a large set of data and examples for training related to the field or the topic. Machine learning can understand the difference among different information only if documents about different topics and information are uploaded during the training process.
- Obtain good results only if the training is frequent (and if the data set grows). Machine learning can improve its knowledge only by adding—over and over again—more information.
- Need different patterns. Too much data of the same **genre** makes the system less accurate. Machine learning can distinguish between the different meanings of the same word, or politics from **ecology**, for example, only if these meanings, or other topics like history, medicine, math, etc., are known by the system.

- Do not learn in real time. You can't add a new concept among the options that machine learning offers.

9 So, you can't hope to train machine learning to identify many different words and different pieces of information without sufficient models and training. Even an extensive knowledge base cannot help you deal with a new word if machine learning has never seen it during training. That means, going beyond text analytics, machine learning offers many opportunities in a variety of fields, and really learns complex information possible.

10 In general, we could say that all applications of machine learning are neither magic boxes nor useless solutions. They cover a very broad range of fields, some very critical fields (for example, life science and health applications)…But how far can they go? Will machines ever learn everything automatically?

Words and Expressions

conversely	/ˈkɒnvɜːsli/	*adv.*	in a way that is the opposite of something 相反地
alien	/ˈeɪliən/	*n.*	a form of life assumed to exist outside the Earth 外星人
classify…as			to decide which type or group somebody/something belongs to 把……归类为
measurement	/ˈmeʒəmənt/	*n.*	a result, usually expressed in numbers, that you obtain by measuring something 测量；尺寸
pixel	/ˈpɪksl/	*n.*	the smallest discrete component of an image or picture on a CRT screen (usually a colored dot) （显示器或电视机图像的）像素
character	/ˈkærəktə(r)/	*n.*	a written symbol that is used to represent speech 字符
plane	/pleɪn/	*n.*	any flat or level surface, or an imaginary flat surface through or joining material objects 平面
prognostic	/prɒgˈnɒstɪk/	*adj.*	connected with the process of making a medical judgment about the likely development of a disease or an illness 预后的；医学预测的
parameter	/pəˈræmɪtə(r)/	*n.*	a factor or limit that affects the way something can be done or made 参数；界限
extraction	/ɪkˈstrækʃn/	*n.*	the process of obtaining something from a mixture or compound by chemical or physical or mechanical means 提取；取出；抽出

therapy	/ˈθerəpi/	*n.*	(medicine) the act of caring for someone (as by medication, remedial training, etc.) 治疗，疗法
regularity	/ˌregjuˈlærəti/	*n.*	the fact that the same thing happens again and again, and usually with the same length of time between each time it happens 规律性
appropriately	/əˈprəʊpriətli/	*adj.*	properly, suitably 适当地；合适地
typically	/ˈtɪpɪkli/	*adv.*	used to say that something usually happens in the way that you are stating 通常，一般
symptom	/ˈsɪmptəm/	*n.*	(medicine) any sensation or change in bodily function that is experienced by a patient and is associated with a particular disease 症状，征兆
narrow down			to reduce the number of possibilities or choices 缩小；减少
inflict	/ɪnˈflɪkt/	*v.*	to make somebody suffer from something unpleasant 使遭受（损伤、痛苦等）
disburse	/dɪsˈbɜːs/	*v.*	to pay out an amount of money usually from a fund which has been collected for a particular purpose 支付，支出
retail	/ˈriːteɪl/	*n.*	the selling of goods to consumers, usually in small quantities and not for resale 零售
sophisticated	/səˈfɪstɪkeɪtɪd/	*adj.*	clever and complicated in the way that it works or is presented 复杂精妙的；先进的
manually	/ˈmænjuəli/	*adv.*	by hand 手动地，用手地
genre	/ˈʒɒnrə/	*n.*	a particular type or style of literature, art, film, or music that you can recognize because of its special features （文学、艺术、电影或音乐的）体裁，类型
ecology	/iˈkɒlədʒi/	*n.*	the relation of plants and living creatures to each other and to their environment; the study of it 生态；生态学

Useful Terms

speech recognition	语音识别
automatic speech recognition	自动语音识别
computer speech recognition	计算机语音识别
speech to text	语音到文本

call routing	呼叫路由选择
domestic appliance	家用电器
intensive care unit	特护病房
independent variable	自变量
information extraction	信息提取
structured information	结构化信息
unstructured information	非结构化信息

Exercises

Comprehension Check

I. Identify the paragraph from which the information is derived.

1. ______________ Classification helps specialists to use measurements of an object to recognize its category.
2. ______________ There is a limit to machine learning in making a possible level of improvement in analyzing unstructured information.
3. ______________ Machine learning enables business to take required decision on time.
4. ______________ A knowledge base cannot offer help if machine learning has never seen new words in training.
5. ______________ The doctors narrow down the disease inflicting the patient according to the results of measurements made by machine learning applications.
6. ______________ Classification is to classify the population under study into different types.
7. ______________ Machine learning applications can be taught by humans the differences between topics, words, and concepts.
8. ______________ Some specialists believe applications of machine learning are both useful and useless.
9. ______________ Machine learning can be used for predicting disease progression, for extracting medical knowledge, for planning and supporting therapy, and for managing patient.

10. ______________ Machine learning needs to be trained sufficiently to identify different words and different pieces of information.

II. Fill in the table below to summarize the applications and their functions of machine learning according to the text. An example has been given.

No.	Applications	Functions
e.g.	*Image recognition*	*Face detection and character recognition*
1.		
2.		
3.		
4.		
5.		

Vocabulary Building

III. Fill in the following blanks with the words and phrases given in the box. Change the form if necessary.

image	text	sufficient	disburse	manually
detection	conversely	classify...as	narrow down	upload

1. The machine can recognize handwritten characters and turn them into printed ______________.
2. You can add the fluid to the powder, or, ______________, the powder to the fluid.
3. The bank agreed further credits to the company because the company can ______________.
4. The ______________ rate for racial crime is appallingly low.
5. The doors can be ______________ operated in the event of fire.
6. The natural ______________ in the poem are meant to be suggestive of realities beyond themselves.
7. What has happened is that the new results ______________ the possibilities.
8. You must choose a file from your hard drive to ______________.
9. We usually ______________ types of characters ______________ good or bad.
10. Lighting levels should be ______________ for photography without flash.

Text C

Basics of Machine Learning

1 I was eating dinner with a couple when they asked what I was working on recently. I replied, "Machine learning." The wife turned to the husband and said, "Honey, what's machine learning?" The husband replied, "Cyberdyne Systems T-800." If you aren't familiar with the Terminator movies, the T-800 is an AI, which has gone very wrong. My friend was a little bit off it. We're not going to attempt to have conversations with computer programs in this book, nor are we going to ask a computer the meaning of life. With machine learning, we can gain insights from a data set; we're going to ask the computer to make some sense from data. This is what we mean by learning, not **cyborg** **rote memorization**, and not the creation of sentient beings.

n. 半机器人
死记硬背

2 Machine learning is actively being used today, perhaps in many more places than you'd expect. Here's a **hypothetical** day and the many times you'll encounter machine learning: You realize it's your friend's birthday and want to send her a card via snail mail. You search for funny cards, and the search engine shows you the ten most relevant links. You click the second link; the search engine learns from this. Next, you check some e-mails, and without you noticing it, the **spam** filter catches unsolicited ads for **pharmaceuticals** and places them in the Spam folder. Next, you head to the store to buy the birthday card. When you're shopping for the card, you pick up some **diapers** for your friend's child. When you get to the checkout and purchase the items, the human operating the cash register hands you a **coupon** for $1 off a six-pack of beer. The cash register's software generated this coupon for you because people who buy diapers also tend to buy beer. You send the birthday card to your friend, and a machine at the post office recognizes your handwriting to direct the mail to the proper delivery truck. Next, you go to the loan agent and ask them if you are **eligible**

adj. 假设的
n. 垃圾邮件
n. 药物
n. 尿不湿
n. 优惠券
adj. 有资格的

for loan; they don't answer but **plug** some financial information about you into the computer and a decision is made. Finally, you head to the **casino** for some late-night entertainment, and as you walk in the door, the person walking in behind you gets approached by security seemingly out of nowhere. They tell him, "Sorry, Mr. Thorp, we're going to have to ask you to leave the casino. Card counters aren't welcome here."

plug *v.* 输入
casino *n.* 赌场

3 In all of the previously mentioned **scenarios**, machine learning was present. Companies are using it to improve business decisions, increase productivity, detect disease, forecast weather, and do many more things. With the exponential growth of technology, we not only need better tools to understand the data we currently have, but also need to prepare ourselves for the data we will have.

scenario *n.* 场景

4 Are you ready for machine learning? In this chapter you'll find out what machine learning is, where it's already being used around you, and how it might help you in the future. Next, we'll talk about some common approaches to solving problems with machine learning. Then you'll find out why Python is so great and why it's a great language for machine learning. Last, we'll go through a really quick example using a module for Python called NumPy, which allows you to do abstract and **matrix** calculations.

matrix *adj.* 矩阵的

5 In all but the most **trivial** cases, insights or knowledge you're trying to get out of the raw data won't be obvious from looking at the data. For example, in detecting spam e-mails, looking for the **occurrence** of a single word may not be very helpful. But looking at the occurrence of certain words used together, combined with the length of the e-mail and other factors, you could get a much clearer picture of whether the e-mail is spam or not. Machine learning is turning data into information.

trivial *adj.* 琐碎的
occurrence *n.* 出现

6 Machine learning lies at the **intersection** of computer science, engineering, and statistics and often appears in other disciplines. As you'll see later, it can be applied to many fields from politics to **geosciences**. It's a tool that can be applied to many problems. Any field that needs to interpret and act on data can benefit from machine learning techniques.

intersection *n.* 交叉
geoscience *n.* 地球科学

7 Machine learning uses statistics. To most people, statistics is an **esoteric** subject used for companies to lie about how great their

esoteric *adj.* 深奥的

products are. (There's a great manual on how to do this called *How to Lie with Statistics* by Darrell Huff. Ironically, this is the best-selling statistics book of all time.) So why do the rest of us need statistics? The practice of engineering is applying science to solve a problem. In engineering we're used to solving a **deterministic** problem where our solution solves the problem all the time. If we're asked to write software to control a **vending** machine, it had better work all the time, regardless of the money entered or the buttons pressed. There are many problems where the solution isn't deterministic. That is, we don't know enough about the problem or don't have enough computing power to properly **model** the problem. For these problems we need statistics. For example, the motivation of humans is a problem that is currently too difficult to model.

deterministic *adj.* 确定性的
vending *adj.* 售卖的
model *v.* 复制

8 In the social sciences, being right 60% of the time is considered successful. If we can predict the way people will behave 60% of the time, we're doing well. How can this be? Shouldn't we be right all the time? If we're not right all the time, doesn't that mean we're doing something wrong?

9 Let me give you an example to illustrate the problem of not being able to fully model the problem. Do humans not act to maximize their own happiness? Can't we just predict the outcome of events involving humans based on this assumption? Perhaps, but it's difficult to define what makes everyone happy, because this may differ greatly from one person to the next. So even if our assumptions are correct about people maximizing their own happiness, the definition of happiness is too complex to model. There are many other examples outside human behavior that we can't currently model deterministically. For these problems we need to use some tools from statistics.

10 In the last half of the 20th century, the majority of the workforce in the developed world has moved from manual labor to what is known as knowledge work. The clear definitions of "moving this from here to there" and "putting a hole in this" are gone. Things are much more **ambiguous** now; job assignments such as "maximizing profits", "minimizing risk", and "finding the best marketing strategy" are all too common. The **fire hose** of information available to us from the World Wide Web makes the jobs of knowledge workers even harder.

ambiguous *adj.* 模糊的
fire hose 消防水管

Making sense of all the data with our job in mind is becoming a more essential skill. With so much of the economic activity dependent on information, you can't afford to be lost in the data. Machine learning will help you get through all the data and extract some information.

Exercises

Comprehension Check

I. Answer the following questions according to the text.

1. Why did the author think his friend's answer to machine learning was a bit wrong?
2. How can the loan agent decide whether customers are eligible for loan?
3. How can machine learning help you detect spam e-mails?
4. Why do we need machine learning to solve statistical problems?
5. What will be a more essential skill for workforce in the future when information is available from World Wide Web?

II. Read the following statements carefully and decide whether they are true (T) or false (F) without turning back to check the text.

1. ______________ Machine learning means machines learn how to memorize information.
2. ______________ You may encounter machine learning many times in a single day.
3. ______________ When you check your e-mail, your filter would recommend you various ads and put them in your mailbox without you noticing them.
4. ______________ The author illustrates some examples of machine learning in action today, like face recognition, handwriting digit recognition, spam filtering in e-mails, and product recommendations.
5. ______________ When the author bought some diapers, the cash register handed him a coupon for $1 off a six-pack of beer, because the store had a beer product promotion.

III. Discuss the following questions based on your understanding of machine learning.

1. What challenges is machine learning facing now?

2. Under the policies of the Chinese government to become the world's primary AI innovation center by 2030, what is China's current situation in gathering machine learning talents?

3. How can a company get over barriers to adopt machine learning?

Writing Skills

同位语

同位语（appositive）指一个名词（或其他形式）对另一个名词或代词进行解释或补充说明。同位语通常与它限定的词连在一起。同位语是由两个或两个以上同一层次的语言单位组成的结构，其中前者与后者所指相同，句法功能也相同，后者是前者的同位语，常用逗号连接。

例 1：Arthur Samuel, an American pioneer in the field of computer gaming and artificial intelligence, coined the term "machine learning".

除了单词、词组以外，也可以使用句子充当同位语从句，用来对其前面的抽象名词进行解释与说明。这类名词有 answer、belief、certainty、conclusion、decision、desire、doubt、dream、evidence、fact、fear、feeling、hope、idea、information、instruction、message、news、order、possibility、problem、promise、question、remark、reply、report、rule、rumor、suggestion、thought、truth、warning 等。

例 2：The users have no idea how long this kind of device will be operating.

引导同位语从句的常用连词有 that、whether、which、who、how、when、where 等。

例 3：The fact is clear that machine learning provides systems with the ability to automatically learn and improve from experience without being explicitly programmed.

Change the following sentences into compound sentences with appositive clauses.

1. Many more companies would benefit from machine learning applications to learn about users and their situations. / There is no doubt about it.

____

____

2. Is machine learning useful? / This debate remains open.

____

____

3. Machine learning algorithms are being developed. / We all know it.

____

____

4. How long will this kind of device be operating? / Even the technicians have no answer to this question.

____

____

5. A Uber self-driving car collided with a pedestrian. / How do you explain it?

____

____

由于英语和汉语具有不同的表达习惯与思维着眼点，两者在词性方面产生了不同的使用方式和使用习惯。例如，英语中有大量使用名词作为中心词的句子，也就是说，英语中有大量的静态句子，这些句子在翻译成汉语时必须加以动态化。为了使译文流畅易懂，并符合目的语的表达方式，英汉互译时常有必要改变原词的词性。词性转换（conversion of part of speech）是英汉互译时最常用的一种变通手法，通过词性的转换可以突破原文的句式，引起句法的转换。词性转换也是避免出现“中式英语”或“英式汉语”的重要手段。译者应根据上下文灵活处理，以求译文符合目的语的表达习惯。

1. 词性转换方法

1）转换为动词

英语中由动词派生的名词和具有动作意义的名词可以转换成汉语动词。介词、形容词、副词等也可以转换为动词。

例 1：Rockets have found application for the exploration of the universe.
火箭已经用来探索宇宙。

例 2：By radar people can see things beyond the visibility of them.
利用雷达，人们能看到视线以外的物体。

例 3：The amount of work is dependent on the applied force and the distance the body is moved.
功的大小取决于施加的力和物体移动的距离。

例 4：Their experiment is over.
他们的试验已经结束了。

2）转换为名词

英语中有很多由名词派生的动词，以及由名词转用的动词。当在汉语中找不到相应的动词时，可以将其转译成汉语名词。形容词、代词等也可以转换为名词。

例 5：The development of scientific research in China is characterized by the integration of theory with practice.
中国科学研究发展的特点是理论联系实际。

例 6：The steam turbine is less economical at cruising speed.
汽轮机的巡航经济性较差。

例 7：The most common acceleration is that of freely falling bodies.
最普通的加速度是自由落体加速度。

3）转换为形容词

当英语中有些抽象名词前加不定冠词用作表语时，常译作汉语中的形容词。副词也可以转换为形容词。

例 8：The maiden voyage of the newly-built steamship was a success.
那艘新建轮船的处女航是成功的。

例 9：The wide application of electronic computers affects tremendously the development of science and technology.
电子计算机的广泛应用，对科学技术的发展有极大的影响。

4）转换为副词

当英语中名词在翻译过程中转换为动词后，原来修饰名词的形容词要相应地转换成副词。

例 10：We must make full use of existing technical equipment.

我们必须充分地利用现有技术设备。

2. 词性转换时注意的问题

词性转换时，需要注意以下三点：首先，需要切合原文的意思，不能为了方便翻译而背离原文意思；其次，翻译时需要与上下文联系起来，做到能让读者清晰、详尽地阅读；最后，翻译时需要确保翻译的准确与顺畅，这是翻译的基本原则与要求。

Translation at Sentence Level

I. Translate the following sentences into Chinese.

1. The dependence of the rate of evaporation of a liquid on temperature is enormous.

____

____

2. The construction of such satellites has now been realized, its realization being supported with all the achievements of modern science.

____

____

3. The flow of electrons is from the negative zinc plate to the positive copper plate.

____

____

4. Attempts to program a computer to integrate gesture with speech began back in the 1970s at the Massachusetts Institute of Technology Architecture Machine Lab.

____

____

5. With the click of a mouse, information from the other end of the globe will be transported to your computer screen at the dizzying speed of seven-and-a-half times around the Earth per second.

____

____

II. Translate the following sentences into English.

1. 我们的河流被污染了，我们的海洋被污染了，我们的自然环境被破坏了。

____

____

2. 现在中国人热衷于从西方国家进口现代化医疗技术，而越来越多的西方人则热衷于接受传统的中医疗法。

____

____

3. 热能是一种能量形式，其他形式的能量都能转化为热能。

____

____

4. 在发达国家，机动车辆是空气污染的主要来源；在发展中国家，这种污染也将增加。

____

____

5. 许多实验室正在研制治疗艾滋病的药物。

____

____

Translation at Paragraph Level

English to Chinese Translation

①Today, announcements about ethical guidelines, principles, and recommendations for AI have been made by governments from many countries, research organizations, and companies. ②To enforce these principles in current AI systems and products, it is vital for the development of governance technology for AI, including federated learning, AI interpretation, rigorous AI safety testing and verification, and AI ethical evaluation. ③These techniques are still under development and are not yet mature enough for widespread commercial adoption. ④Major technical obstacles are deeply rooted in fundamental challenges for modern AI research, such as human-level moral cognition, common-sense ethical reasoning, and multidisciplinary AI ethics engineering.

本段一共四句话。第一句是以名词 announcement 作为中心词的被动句，主干部分为 announcements have been made by…。英译汉时，为了符合汉语动态化的表达习惯，首先需要把被动句改为主动句，接着把 announcement 的词性改为动词。如果不进行词性转换，那么这句话可能被译为："今天，许多国家的政府、研究机构和公司对关于人工智能的道德准则、原则和建议进行了宣布"，显得生硬。词性转换后，这句话可以译为："今天，许多国家的政府、研究机构和公司都宣布了他们关于人工智能的道德准则、原则和建议。"

第二句是以 development 为中心词的静态句子，主干部分是 it is vital for the development of…。为了符合汉语动态化的表达习惯，将 development 译成动词更为合适。如果不进行词性转换，这句话可能被译为："要在当前的人工智能系统和产品中执行这些原则，关于人工智能治理技术的开发至关重要，包括联合学习、人工智能翻译、严格的人工智能安全检测和认证，以及人工智能伦理评估"，显得生硬。词性转换后，这句话可以译为："要在当前的人工智能系统和产品中执行这些原则，开发人工智能治理技术至关重要，包括联合学习、人工智能翻译、严格的人工智能安全检测和认证，以及人工智能伦理评估。"

第三句中的 development 和 mature 同样可以转换成动词，译为："这些技术仍在开发中，尚未成熟到可以广泛应用于商业领域。"第四句的主干部分是"Major technical obstacles are deeply rooted in fundamental challenges."，谓语 deeply rooted in 的本意是"深深植根于"。这句话把重大技术难题看成了挑战，而具体的重大技术难题在句尾也进行了进一步的解释与说明。所以在译成汉语时，句尾的解释说明可以和 Major technical obstacles 放在一起作主语。这句话可以译为："人类层面的道德认知、常识伦理推理、多学科的人工智能伦理工程等重大技术难题是现代人工智能研究面临的主要挑战。"

III. Translate the following paragraph into Chinese.

Ethics and governance are vital to the healthy and sustainable development of artificial intelligence. With the long-term goal of keeping AI beneficial to human society, governments, research organizations, and companies in China have announced ethical guidelines and principles for AI, and have launched projects of the development of AI governance technologies.

____

____

____

____

____

____

Chinese to English Translation

①人工智能的迅速发展预示着我们的社会即将发生根本性的变革。②这一转变可以成为构建人类社会共享未来、促进社会和自然环境可持续发展的大好机会。③但如果没有有效的治理和监管，它可能会带来前所未有的消极影响。④为了确保这些变化在完全融入我们日常生活的基础设施之前是有益的，我们需要建立一个坚实可行的人工智能治理框架，根据人类的伦理和价值观来规范人工智能的发展。

本段一共四句话。第一句的主干部分清晰明了，即“迅速发展预示着变革”。为了让整个句子的结构紧凑不松散，可以把动词“即将发生”转换成形容词，和“根本性的”一起作“变革”的修饰语。这句话可以译为：“The rapid development of AI indicates an upcoming fundamental transformation of our society. ”。

第二句的主干为“这一转变成为大好机会”。但是需要注意“大好机会”的定语部分“构建人类社会共享未来、促进社会和自然环境可持续发展”的翻译方法。第一，汉语中的前置定语译成英语时，通常作后置定语，避免头重脚轻。第二，这部分定语共有四个动词：“构建”“共享”“促进”和“发展”。如果把它们都翻译成英语动词，那么句子表达未免乏味，句子结构也较为松散，因此具体的处理方法是：“构建”和“促进”引领两个并列的不定式动词短语，词性保持不变；“共享”和“发展”进行词性转换，分别译成形容词和名词。整个句子可以译为：“This transformation can be a great opportunity to construct a human community with a shared future, and to promote the sustainable development of society and the natural environment.”。

第三句的结构是“如果没有……（那么）可能……”，可以译成典型的由 without 引导的条件句。without 后面要跟名词或名词短语，所以“治理”和“监管”两个动词可以转换成名词。这句话可以译为：“But without sufficient and effective governance and regulation, its implications might be unprecedented and negative.”。

第四句的主干部分是“为了确保……是有益的，我们需要建立框架，来规范人工智能的发展”，介词短语“根据人类的伦理和价值观”可以放在句首或句尾。这句话可以译为：“In order to ensure that these changes are beneficial before they are completely embedded into the infrastructure of our daily life, we need to build a solid and feasible AI governance framework to regulate the development of AI according to the ethics and values of humanity.”。

IV. Translate the following paragraph into English.

机器学习可以用来提取信息。具体来说，它从非结构化数据，如网页、文章、博客、业务报告和电子邮件等中提取结构化信息，利用相关数据库存放提取的信息，以便计算机算法后续进行查找。如今，信息提取正在成为大数据行业发展的关键。

____

____

____

____

____

____

Workshop

I. Choose one domain that you think needs machine learning technology most in the future and share your reasons. Then make a five-minute oral presentation to the class.

II. Fill in the table below to make a comparison between machine learning and human learning and discuss with your partners about which one is better.

Details	Machine Learning	Human Learning
Advantages		
Disadvantages		

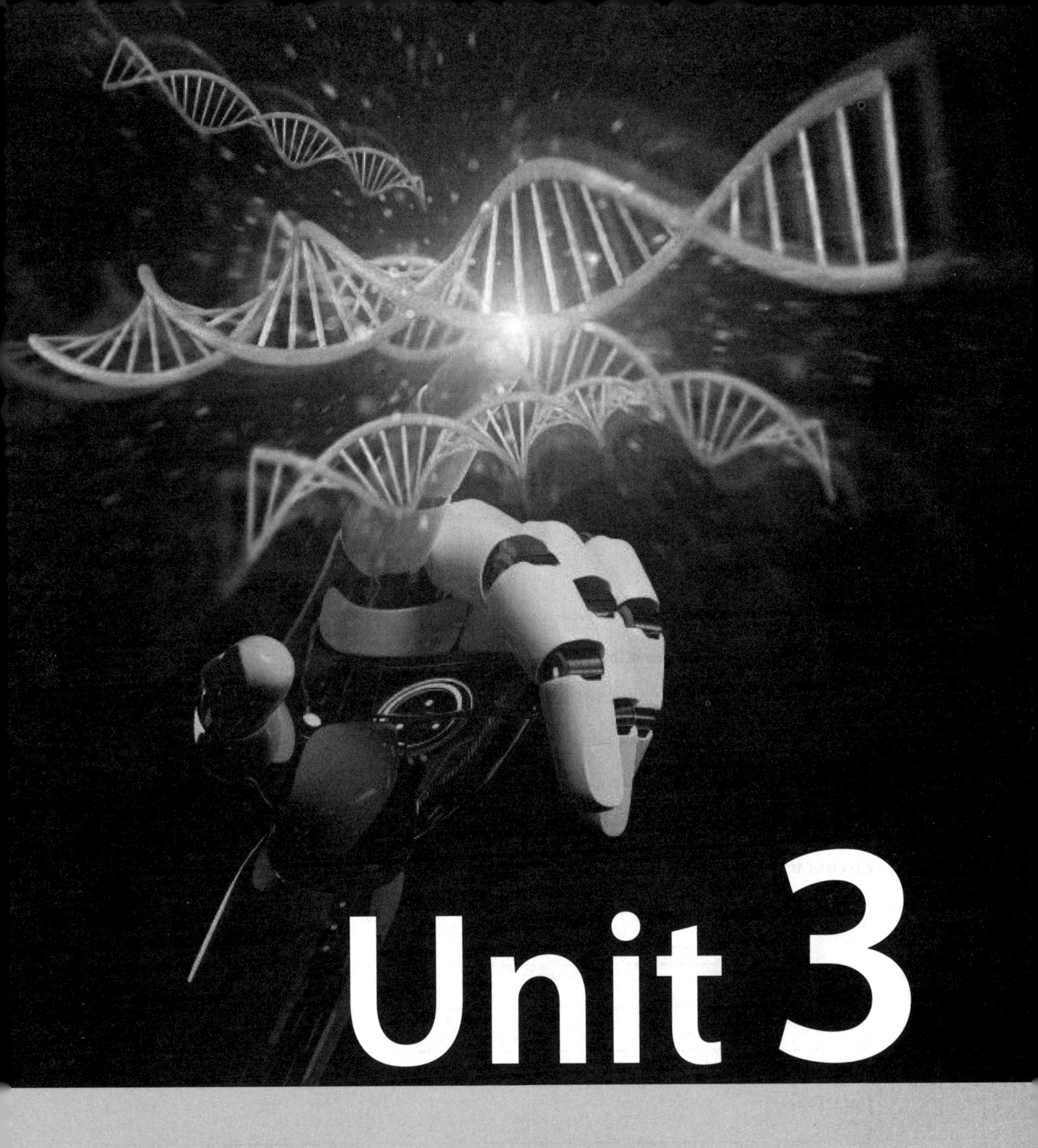

Unit 3

Artificial Intelligence and Big Data

Learning Objectives

In this unit, you will learn:

- a brief introduction to big data;
- the combination of big data and artificial intelligence;
- the analysis techniques of big data;
- writing skills—paragraph development;
- translating skills—adjustment of word order.

Lead-in

I. Discuss the following questions with your partners.

1. How much do you know about big data?
2. Why is big data important nowadays?
3. What is the relationship between big data and artificial intelligence?

II. Work in pairs to find out three fields where big data is frequently collected and discuss what type of data is needed there.

Fields:

1. ______________________________
2. ______________________________
3. ______________________________

Data Types:

1. ______________________________
2. ______________________________
3. ______________________________

An Introduction to Big Data

1 According to the recent executive survey of Fortune 1,000 corporations, 97.2% of executives report that their companies are investing in building or launching big data and AI **initiatives**. Due to the faster growth and greater **availability** of data, AI and big data are closely connected with each other in many ways, enabling companies to develop AI and cognitive initiatives within their organizations.

2 Among the survey participants, most of them are the executives from data-intensive industries, especially financial services companies, which **constituted** 77.2% of the survey respondents. Financial services companies have long been at the forefront of industry because they not only maintain large amounts of **transactional** and customer data, but also have developed **robust** data management and data governance processes for decades. Industries such as life sciences possess vast storage of scientific and patient data that have not yet been used.

3 Data has **swept into** every industry and business function. The scale of data sets keeps changing and varies from organization to organization. Dealing with the large data sets has far exceeded the capability of traditional computers. The non-traditional strategies and technologies are needed to gather, organize, and process information from large data sets.

What Is Big Data?

4 An exact definition of "big data" is difficult to **nail down** because projects, **vendors**, practitioners, and business professionals use it quite differently. Generally speaking, big data is a term to describe the large data sets collected by firms and governments on a day-to-day basis. This term is also applied to technologies and strategies to work with this type of data.

5 In 2001, Gartner's Douglas Laney first used "three Vs of big data" to describe the main characteristics of big data, explaining why big data differs **significantly** from other data processing.

Volume

6 The amount of data that's being created and stored on a global level keeps growing. The data sets are so large that they cannot be processed or stored with traditional tooling or on a single computer. This becomes a challenge of pooling, **allocating**, and **coordinating** resources from

groups of computers. Cluster management and algorithms can ease the burden by breaking tasks into smaller pieces.

Velocity

7 Another significant feature that makes big data different from other data systems is the speed that information moves through the system. Data is frequently flowing into the system from different sources and is often expected to be dealt with in a timely manner. Data is constantly being added, processed, and analyzed in order to pick up valuable information early when it is most relevant. These ideas require robust systems to guard against failures along the data pipeline.

Variety

8 Big data problems are often unique because of the wide range of both sources and quality.

9 Data can come from a variety of sources, like application and server logs, social media, physical device sensors, and other providers. Big data seeks to handle possibly useful data **regardless of** where it's coming from. The formats and types of media can vary significantly as well. Apart from text files and structured logs, there are also rich media like images, video files, audio recordings, etc. While more traditional data processing systems might expect data to enter the system already labeled, formatted, and organized, big data systems usually accept and store data closer to its **raw** state.

What Does a Big Data Life Cycle Look like?

10 So how is data actually processed in a big data system? The following steps might not be true in all cases, but they are widely used. The general categories of activities involved with big data processing are as follows:

Ingesting Data into the System

11 Data ingestion is the process of taking raw data and adding it to the system. During the ingestion process, some level of analysis, sorting, and labeling usually takes place. The captured data should be kept as raw as possible so that greater **flexibility** is possible in the further process.

Persisting the Data in Storage

12 The volume and availability of incoming data, and the **distributed** computing layer require more complex storage systems. This ensures that the data can be stored by compute resources, can be put into the RAM for in-memory operations, and can well handle component failures.

Computing and Analyzing Data

13 Once the data is available, the system can begin processing the data. Batch processing is one method of computing over a large data set. The process involves breaking work up into smaller pieces, scheduling each piece on an individual machine, regrouping the data based on the intermediate results, and then calculating and **assembling** the final result.

14 Another method is real-time processing. Real-time processing demands that information be processed and made ready immediately and requires the system to react immediately when new information becomes available.

Visualizing the Result

15 Recognizing trends or changes in data over time is often more important than the values themselves. Visualizing data is one of the most useful ways to spot trends and make sense of a large number of data points.

16 In conclusion, many firms and organizations are turning to big data for certain types of workloads and using it to supplement their existing analysis and business tools. Big data systems are uniquely suited for discovering difficult-to-detect problems and providing insights into behaviors that are impossible to find through traditional methods.

Notes

Fortune 1,000 a reference to a list maintained by the American business magazine *Fortune*. The list is of the 1,000 largest American companies, ranked by revenues. The list draws the attention of business readers seeking to learn the influential players in the American economy and prospective sales targets, as these companies tend to have large budgets and staff needs.

Gartner officially known as Gartner, Inc., a global research and advisory firm providing insights, advice, and tools for leaders in IT, finance, HR, customer service and support, legal compliance, marketing, sales, and supply chain areas across the world.

Words and Expressions

initiative	/ɪ'nɪʃɪətɪv/	*n.*	an important new plan or process for achieving a particular aim or for solving a particular problem 计划；措施
availability	/əˌveɪlə'bɪləti/	*n.*	the quality of being at hand when needed 可用性，可获得性
constitute	/'kɒnstɪtjuːt/	*v.*	to be the parts that together form something 组成，构成
transactional	/træn'zækʃənl/	*adj.*	relating to an act of buying or selling 交易的
robust	/rəʊ'bʌst/	*adj.*	strong and not likely to fail or become weak 强劲的；有活力的

sweep into			to rush into 涌入
nail down			to define clearly 确定
vendor	/ˈvendə(r)/	*n.*	a company that sells a particular product 销售商
significantly	/sɪɡˈnɪfɪkəntli/	*adv.*	in an important way or to an important degree 显著地，明显地
allocate	/ˈæləkeɪt/	*v.*	to give something officially to somebody/something for a particular purpose 分配；拨给
coordinate	/kəʊˈɔːdɪneɪt/	*v.*	to organize an activity so that the people involved in it work well together and achieve a good result 协调（多人参加的活动）
velocity	/vəˈlɒsəti/	*n.*	the speed of something that is moving in a particular direction 速度
regardless of			without taking into account 不管，不顾
raw	/rɔː/	*adj.*	not yet organized into a form in which it can be easily used or understood 未经处理的；未经分析的；原始的
ingest	/ɪnˈdʒest/	*v.*	to absorb information, knowledge, etc. 摄入；吸收
flexibility	/ˌfleksəˈbɪləti/	*n.*	the ability to change or be changed easily to suit a new situation or condition 灵活性
persist	/pəˈsɪst/	*v.*	to continue to exist or happen 继续存在（发生）
distribute	/dɪˈstrɪbjuːt/	*v.*	to spread something or different parts of something over an area 使分布；使散开
assemble	/əˈsembl/	*v.*	to bring people or things together as a group 收集；聚集

Useful Terms

RAM	随机存取存储器
batch processing	成批处理

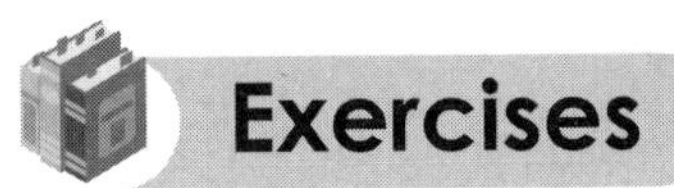

Exercises

Comprehension Check

I. Answer the following questions according to the text.

1. Why have financial services companies long been at the forefront of industry?
2. What kind of data does life sciences industry possess?
3. What do "three Vs of big data" refer to?
4. What activities are involved with big data processing?
5. Why is visualizing data important?

II. Read the following statements carefully and decide whether they are true (T) or false (F) without turning back to check the text.

1. ____________ A vast majority of Fortune 1,000 corporations are investing in setting up big data and AI initiatives.
2. ____________ The scale of big data sets may be different from organization to organization.
3. ____________ Big data systems only handle useful data from internal systems.
4. ____________ Sorting and labeling data usually takes place at the third step of a big data life cycle.
5. ____________ RAM is used to ingest data into the big data system.

III. Choose the best answer to each of the following questions according to the text.

1. What enables companies to build AI initiatives within their organizations?

   A. The increasing computing power or storage of a single computer.

   B. The vast investment from Fortune 1,000 corporations.

   C. The fast growth and greater availability of data.

   D. The expanding scale of traditional businesses.

2. Which industry is playing a leading role in building big data and AI initiatives?

   A. Social media.

   B. Life sciences.

   C. Computer sciences.

D. Financial services.

3. Which is NOT the characteristic of big data according to Douglas Laney?

A. The wide range of sources and quality.

B. The long life cycle of the system.

C. The fast speed that information moves through the system.

D. The much larger data sets to be processed or stored.

4. What kind of data is usually accepted by a big data system?

A. Labeled data.

B. Formatted data.

C. Raw data.

D. Organized data.

5. What does real-time processing refer to?

A. It refers to process the information after picking out the really useful data.

B. It refers to process the information immediately when it is available.

C. It refers to process the information after waiting for a period of time.

D. It refers to process the information when it is worth analyzing.

Vocabulary Building

IV. Fill in the following blanks with the words and phrases given in the box. Change the form if necessary.

constitute	sweep into	assemble	regardless of	transactional
flexibility	initiative	capability	raw	significantly

1. If an enterprise fails to grasp the ______________ in the market, it will lose the leading position.

2. The new regulation gives employees more ______________ in the workplace.

3. As foreign brands ______________ the domestic market, Chinese home appliance enterprises are facing great competition.

4. Supermarkets have a lot of ______________ data every day, which helps to know consumers' preferences.

5. Female employees ______________ less than 5% of the total number of the staff.

6. Training methods used by the new football coach differ ______________ from those used by older ones.

7. The software is designed to ______________ test data immediately after students submit answers.

8. The system analyzes and stores information in real time, ______________ the arrival time of the message.

9. They both have the ______________ of finishing the task independently.

10. The factory buys its ______________ materials cheaply elsewhere and sells its finished products at a higher price.

V. Match the words in the left column with the explanations in the right column.

1.	coordinate	A.	to absorb information, knowledge, etc.
2.	vendor	B.	relating to the mental process involved in knowing, learning, and understanding things
3.	ingest	C.	a company that sells a particular product
4.	cognitive	D.	an important new plan or process for achieving a particular aim or for solving a particular problem
5.	initiative	E.	to organize an activity so that the people involved in it work well together and achieve a good result

Word Formation

VI. Fill in the following blanks with the words in capitals. Change the form if necessary. An example has been given.

e.g. *There has been a significant development of artificial intelligence in the past few years.* **DEVELOP**

1. We need to reach an ______________ on the future cooperation. **AGREE**

2. Have you asked the technician to check the new ______________ before we use it? **EQUIP**

3. How would you evaluate the return on ______________ of driverless cars? **INVEST**

4. They are discussing how to avoid the risk of ______________ in cloud computing. **MANAGE**

5. Variations of the product packaging can be arranged to suit your ______________. **REQUIRE**

Translation

VII. Translate the following sentences into English with the words and phrases in brackets.

1. 大型数据集的规模已经远远超过了传统计算机的处理能力。(deal with; exceed)

____

____

2. 大数据与其他数据系统的另一个显著不同之处在于信息在系统中的传播速度。(differ from; move through)

____

____

3. 大数据将所有信息统一到一个系统中来处理可能有用的数据，而不管这些数据来自何处。(handle; regardless of)

____

____

Text B

Merging Big Data with Artificial Intelligence Is the Next Step

1 Artificial intelligence is one of the hottest trends in technology right now. Big data is another fashionable technology. What will happen if we **merge** big data with AI? Researchers want to find out the ways to unite big data with AI. We have only recently realized how powerful big data can be, and through the **combination** with AI, big data is rapidly moving to a level of **maturity** that will lead to a even greater, industry-wide disruption.

2 The application of artificial intelligence to big data is probably the most important modern breakthrough of our time. With the help of data, businesses create more value than before. The availability of big data has led to **unprecedented** breakthroughs in machine learning that were previously impossible.

3 Being able to obtain large volumes of data sets, businesses can get meaningful learning and **come up with** amazing results. No wonder companies are rapidly shifting from a **hypothesis**-based approach to a more focused "data first" strategy. But how does big data drive rapid breakthroughs in artificial intelligence?

4 Businesses can now deal with large volumes of data, which was impossible before because of the limitation of technology. They no longer need to buy expensive hardware and software. The widespread availability of data is the most important model shift, which has promoted **innovation** in the industry. The availability of large data sets and the **emergence** of more complex AI algorithms promote the fast development in machine learning.

5 The best example of these breakthroughs is **virtual** agents (more commonly known as chatbots). Virtual agents have made a deep impression on people. Before that, chatbots had trouble identifying certain phrases or regional accents, dialects or small difference. In fact, most chatbots **get stumped** by the simplest words and expressions, such as mistaking "Queue" for "Q" and so on. With the combination of big data and AI, however, virtual agents can self-learn in an amazing way.

6 A good example of self-learning virtual agents is Amelia, a "cognitive agent" recently developed by IPsoft. Amelia can understand everyday languages and learn them quickly, even becoming smarter over time. She is placed at SEB bank's help desk in northern Europe, along with many public sector agencies. The management's response to Amelia has been very positive.

7 Google is also taking a deeper look at big data-driven AI learning. DeepMind, the artificial intelligence company owned by Google, has developed an AI that can teach itself to "walk, run, jump, and climb without any prior guidance". The AI has never been taught what walking or running was, but through trial and error, it has learned on its own.

8 The **implications** of these breakthroughs in artificial intelligence are amazing and could lay the foundation for further innovations in the future. However, self-taught algorithms can also have terrible consequences, and if you are not too busy to notice this, you have probably observed a lot in the past.

9 Not long ago, Microsoft introduced its own AI chatbot named Tay. The robot was able to chat with the public and could learn through human interactions. However, a day after the bot was introduced to Twitter, Microsoft **pulled the plug** on the project. Tay learns mainly through human interactions, and in less than 24 hours has **transformed** from an **innocent** AI teen girl to an evil, Hitler-loving, impolite, and lying robot.

10 Should the **evolution** of AI concern us? Some fans of science fiction films such as *The Terminator* worry that as AI gains access to big data, it could become "self-aware" and start massive cyberattacks that could even take over the world. More realistically speaking, it may replace human jobs. Judging from the speed of AI learning, we can understand why many people in the world care

about AI's self-learning and the big data access it enjoys. Whatever the case, the **prospects** are both attractive and frightening.

11 It is hard to say how the world will react to the combination of big data and artificial intelligence. However, like everything else, it has its **virtue and vice**. For example, it is true that self-learning AI will predict a new age, where chatbots become more efficient in answering users' complex questions. Perhaps we would finally see AI bots waiting to greet us on banks' help desks. And, through self-learning, the bot will have all the knowledge it may need to answer all our questions in a way unlike any human assistant.

12 Regardless of its application, it is safe to say that the combination of big data with artificial intelligence will predict an era full of new possibilities and shocking breakthroughs and innovations in technology. Let us only hope that the virtue of this union will outweigh the vice.

Notes

IPsoft an American technology company founded in 1998 by Chetan Dube. It is based in New York, with 16 offices in 11 different countries. It has three products: IPcenter, 1Desk, and Amelia.

SEB Skandinaviska Enskilda Banken, a Swedish financial group for corporate customers, institutions, and private individuals with its headquarters in Stockholm. Its activities comprise mainly banking services, but SEB also carries out significant life insurance operations and owns Eurocard.

DeepMind a British artificial intelligence company founded in September 2010. The company is based in London, but has research centers in California, Edmonton, and Montreal. Acquired by Google in 2014, the company has created a neural network that learns how to play video games in a fashion similar to that of humans. The company made headlines in 2016 after its AlphaGo program beat a human professional Go player for the first time.

Words and Expressions

merge /mɜːdʒ/ *v.* to combine two or more things to form a single thing 合并，并入，结合

combination /ˌkɒmbɪˈneɪʃn/ *n.* the act of joining or mixing two or more different things together to form a single unit 结合，联合，混合

maturity /məˈtʃʊərəti/ *n.* the state of being fully developed 成熟

unprecedented	/ʌn'presɪdentɪd/	*adj.*	that has never happened before 史无前例的
come up with			to put forward 提出，想出
hypothesis	/haɪ'pɒθəsɪs/	*n.*	an idea or explanation of something that is based on a few known facts but that has not yet been proved to be correct 假设
innovation	/ˌɪnə'veɪʃn/	*n.*	the introduction of new things, ideas, or ways of doing something 创新，改革
emergence	/ɪ'mɜːdʒəns/	*n.*	the fact of starting to exist or becoming known for the first time 出现，涌现
virtual	/'vɜːtʃuəl/	*adj.*	generated by a computer to simulate real objects and activities （计算机仿真）虚拟的
get stumped			to meet difficulties 遇到困难
implication	/ˌɪmplɪ'keɪʃn/	*n.*	a possible effect or result of an action or a decision 可能的影响（或作用、结果）
pull the plug			to finish 结束，终止业务
transform	/træns'fɔːm/	*v.*	to completely change the appearance, form, or character of somebody/something, especially in a way that improves it 使改变，使改观；使转化
innocent	/'ɪnəsnt/	*adj.*	having little experience of the world, especially of evil or unpleasant things 天真无邪的
evolution	/ˌiːvə'luːʃn/	*n.*	a process of gradual development in a particular situation over a period of time 演变；发展
prospect	/'prɒspekt/	*n.*	an idea of what might or will happen in the future 前景，展望
virtue and vice			advantages and disadvantages 优点和缺点

Useful Term

cyberattack	网络攻击

Exercises

Comprehension Check

I. Identify the paragraph from which the information is derived.

1. ______________ Big data enables businesses to move from a hypothesis-based research approach to a more focused "data first" strategy.
2. ______________ Big data is rapidly developing towards a level of maturity by uniting with AI.
3. ______________ Virtual agents are regarded as the best example of breakthroughs in machine learning.
4. ______________ Amelia is a self-learning virtual agent developed by IPsoft.
5. ______________ Tay is the name of Microsoft's artificial intelligence chatbot.
6. ______________ Artificial intelligence may become "self-aware" and replace human jobs.
7. ______________ Google's DeepMind has developed a self-learning AI.
8. ______________ The availability of big data has led to the great development in machine learning.
9. ______________ The virtual agent Amelia is placed at SEB bank's help desk.
10. ______________ The prospects of self-learning AI are both attractive and terrifying.

II. Fill in the table below to summarize the description of virtual agents according to the text.

Details	Amelia	DeepMind's AI	Tay
Company			
Main Features			
Effects			

Vocabulary Building

III. Fill in the following blanks with the words and phrases given in the box. Change the form if necessary.

combination	implication	come up with	identify	transform
get stumped	evolution	innovation	emergence	merge

1. Newcomers wanted to see the core material, but ______________ by the access rights.
2. Capturing data is useful to ______________ trends or patterns in customer or employee behavior.
3. We must encourage ______________ in products and technology if the company wants to remain competitive.
4. Fierce domestic competition has forced the company to develop new products in ______________ with several overseas partners.
5. Artificial intelligence has become widespread in every industry where decision-making is being fundamentally ______________ by thinking machines.
6. Technicians ______________ some solutions to the problem of inaccurate speech recognition by robots.
7. The process of human ______________ is long and worth studying.
8. The company is cutting back its spending and I wonder what the ______________ are for our department.
9. With the ______________ of various digital devices, people can use computers to process images and enhance the quality of images.
10. After taking data from two different source fields, please ______________ them into a single target field.

Analysis Techniques of Big Data

1 The **abundance** of data presents both opportunities and challenges. An important problem facing researchers is how to use analytical and statistical techniques to analyze data. Statistical methods developed in the 19th and mid-20th centuries can rarely handle very large and complex data sets. The amount and variety of data collected today are so large and fast that traditional methods of data analysis are no longer sufficient to handle the flood of information.

n. 充裕，丰富

2 To deal with these challenges, researchers have developed so-called "**predictive** analytics" or "user behavior analytics" to manage big data. In these analytical approaches, a variety of statistical techniques can be used to extract value from data, including predictive modelling, machine learning, and data mining. The advantage of these methods is that they create learning algorithms that can spot patterns with predictive power. Typically, this involves analyzing trends in historical and transactional data to make new discoveries or predictions about the future or other unknown events.

adj. 预测性的

3 Businesses and governments typically use one of three **computational** models to process large data sets: data mining, artificial neural network (ANN), and machine learning.

adj. 计算的

4 In general, data mining (also known as knowledge discovery in databases) uses existing information and data analysis to search for hidden or emerging patterns in the data to explain a specific phenomenon. Data mining is usually used to classify, sort, and describe data, but it can also be used to describe new discoveries and predict future trends.

5 Artificial neural network method is different from data mining. ANN models are designed to imitate the human brain. Like a brain

made up of billions of neurons, the role of an ANN model is to understand data input in order to make future predictions, reduce information **overload** or noise, and classify events. The ANN program recognizes patterns in the data and then **infers** connections, characteristics, and outputs to make decisions faster and more accurate.

n. 超载
v. 推断

6 This brings us to machine learning. Machine learning models gain knowledge from existing data, which provides the basis for machine self-learning. Machine learning uses algorithms to build models that clarify connections, make **assumptions**, and apply what has been learned to make future predictions. Used properly, machine learning algorithms can provide instant, real-time advice for decision makers.

n. 假定，设想

7 The use of data mining, ANN, and machine learning techniques is not without **controversy**. Whenever companies or governments collect micro-data on consumers and citizens, there are obvious ethical issues. Policymakers and regulators have the right to ask how much information a company or government should be allowed to collect about consumer choices. Questions to be answered include: How should the person collecting the information be allowed to use it? How does the use of big data **violate** consumer privacy? Who should be allowed to access the data?

n. 争论，争议
v. 侵犯

8 In addition to these ethical and privacy concerns, there is another fierce debate going on among economists about the effectiveness of statistical techniques used to deal with big data. As a financial service professional, this leads you to a decision point. Are you willing to have computer models that sort and **synthesize** data in a way that gives insights into the thoughts and behaviors of clients and other consumers?

v. 合成；综合

9 Big data approach is currently being used across a wide range of sectors in the financial services industry, such as insurance, healthcare, retirement planning, risk assessment, telecommunications, retail, and **fraud** detection. The use of big data is also shaping the understanding of other topics in financial services. There are many examples where big data is being used to identify new market segments, create new products and services, and deepen customer engagement.

n. 欺诈

10 Researchers at the International Finance Corporation (IFC) have used big data from mobile network operators and telecommunications

companies to predict future financial business possibilities. The researchers **tracked** the behaviors of four million mobile phone users, including voice, data, and geographic information. They then used the data to reveal how these usage patterns relate to the use of digital financial services (DFS) such as mobile banking. They found that DFS users had a larger number of phone calls, longer call durations, a larger number of text messages, and a larger size of the customers' social network than non-users. They used their big data discovery to create maps that showed the actual distribution of DFS users and non-users in each country, and then identified those locations that had the greatest concentration of potential new DFS users. The use of mobile phone data allowed DFS providers to target more precisely those who were most likely to adopt mobile banking and other digital financial services.

v. 跟踪，追踪

11 As the above brief introduction to big data shows, the use of data analysis techniques such as data mining, ANN, and machine learning will continue to grow in the future. It is reasonable to expect continued debate about the theoretical basis and practical application of these approaches, but in the end, financial services can expect more value from big data.

Exercises

Comprehension Check

I. Answer the following questions according to the text.

1. What is the technical challenge of big data?
2. What computational models are mainly used by enterprises and governments to deal with big data?
3. What are the controversies about the use of statistical techniques?
4. Can you give some examples of big data applied to the financial services industry?
5. What did the IFC researchers find when they tracked the behaviors of mobile phone users?

II. Read the following statements carefully and decide whether they are true (T) or false (F) without turning back to check the text.

1. ____________ Traditional methods of data analysis can hardly deal with the large and complex data sets.
2. ____________ Artificial neural network imitates the function of human brain to process data input and output.
3. ____________ IFC tracked the data on mobile phone users to help governments create maps.
4. ____________ Mobile phone data helps DFS providers target potential users of mobile banking.
5. ____________ Financial services professionals agree on the effectiveness of statistical techniques for processing big data.

III. Discuss the following questions based on your understanding of big data.

1. Data breach is the release of secure information to an untrusted environment. Millions of people are affected by data breach every year. Can you give some examples of data breach?
2. How can people avoid data breach while enjoying the benefits of big data?
3. How does big data affect our daily lives?

Writing Skills

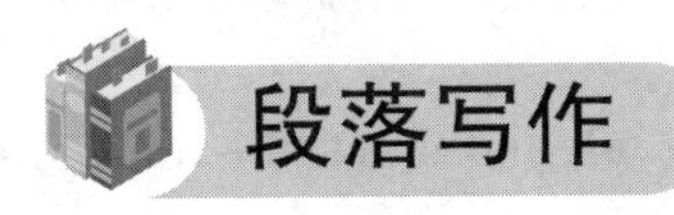

段落写作

在科技应用文中，一个规范的段落通常既有主题句（topic sentence）来表达中心思想，又有扩展句（developing sentence）来围绕中心思想进行阐述或论证，有时还有结尾句（concluding sentence）来总结或强调段落主旨。

主题句是一个段落的核心，它表明作者的观点和态度，同时反映段落的中心思想。主题句需要有明确的观点和一定的概括性。扩展句围绕主题句进行叙述、说明或论述。结尾句用一句话将整段内容进行归纳总结，不仅需要简明扼要，还需要符合主题句的主旨。

例 1：How is data actually processed when dealing with a big data system? First of all, ingest data into the system. Next, persist the data in storage. After that, compute and analyze data. The last step is to visualize the result.

例 1 的主题句是："How is data actually processed when dealing with a big data system?"。之后四个扩展句按照数据处理过程的先后顺序依次展开，并且每句的句首都使用了衔接词：First of all、Next、After that 和 The last step。

例 2：Difficulty in accessing data can prevent new products from entering the market, which has two implications. First, consumers don't benefit from the innovative new products that a potential entrant might offer. Second, established firms, knowing that there is little chance of entry from a potential rival, may not see the need to innovate and improve on their existing products. Without the threat of new products, then, there may be fewer incentives to innovate.

例 2 的主题句是："Difficulty in accessing data can prevent new products from entering the market, which has two implications."。之后两个扩展句按照并列关系依次展开，并且每句的句首都使用了衔接词：First 和 Second。最后的结尾句直接得出结论。

为了使段落里的各句内容连贯、层次分明，需要特别注意句子之间的衔接（cohesion）。一个简便易行的方法就是根据各句之间的逻辑关系使用合适的衔接词。例如：

- 表先后：first(ly), second(ly), to begin with, to start with, first of all, next, finally, last but not least, at last
- 表并列：and, as well as, both…and…, either…or…
- 表转折：but, however, though, despite, in spite of
- 表因果：because, owing to, due to, so, since, therefore, thus, hence
- 表递进：moreover, furthermore, in addition (to), not only…but also…
- 表举例：for example, for instance, such as
- 表对比：in contrast (to), on the contrary, but, however
- 表总结：to sum up, to conclude, in summary, in conclusion, in short

I. Read the following paragraph and mark the topic sentence, developing sentences, and concluding sentence.

Looking back at the history of the Internet, we find that it has had an increasing impact on the world. In the early days of the Internet in 1969, no one could ever have imagined the enormous impact that the Internet has today. Back then, the Internet was used to send e-mail messages between two computers and was designed with U.S. Defense in mind. It was not until the mid-1990s when the Internet started taking off on a massive scale through Web-based services. Since then, the amount of data has exploded with richer and richer content. Today, the rate at which content is improving and expanding is mind-blowing. Therefore, one has to say that the Internet is one of the most important technological breakthroughs of modern society.

II. Suppose you are going to introduce a mobile phone app that you often use within one paragraph. Part of the topic sentence has been given. Please complete the topic sentence and continue to write the developing sentences and concluding sentence.

______________________________ is my favorite mobile app because it has so many functions.

____

____

____

____

____

语序调整

汉语中很多句子为主题—评述句，而绝大多数英语句子为主谓句。汉语重意合，句子由主题语统领，评述语紧随其后。汉语句子还频繁使用动词，并按动作发生的先后顺序或逻辑顺序，逐步交代，层层展开。英语重形合，句子必须严格按照句法结构组合起来。英语首先突出句子主干部分（主语、谓语和宾语），之后再运用各种关系词把各种从句及其他短语与主干部分联系起来。因此在英译汉时，许多英语主谓句要译为汉语主题—评述句。一旦确立主题语和评述语，译文的结构框架就确立了，再根据各语言单位之间的逻辑关系灵活调整语序（adjustment of word order）。了解英语和汉语的语言特点，有助于翻译出高质量的译文。

英语主谓句在译成汉语主题—评述句时，通常采用以下三种语序调整策略：

1. 状语（从句）提到动词前

例 1：Data is frequently flowing into the system from different sources and is often expected to be dealt with in a timely manner.

通常不同来源的数据会流入系统，且经常会得到及时处理。

例 1 是一个由 and 连接的并列句。介宾短语 from different sources 作地点状语，翻译时需要提到动词“流入”之前；介宾短语 in a timely manner 作方式状语，翻译时需要提到动词“处理”之前。

2. 定语（从句）提到先行词前

例 2：Real-time processing is best suited for analyzing smaller chunks of data that are changing rapidly.

实时处理最适合分析快速变化的小块数据。

例 2 中有一个由 that 引导的定语从句，修饰限定先行词 data，翻译时需要调整语序，先译出定语从句的内容，再译出先行词的意思。

3. 用人或有生命的物体作主题语

例 3：There are many reasons why vendors encourage customers to use cashless payment methods.

商贩们鼓励顾客使用无现金支付方式，这有很多原因。

例 3 是个 there be 句型，强调 many reasons，但是在英译汉时需要调整语序，将“商贩”作为主题语前置，随后是评述语。

Translation at Sentence Level

I. Translate the following sentences into Chinese.

1. The widespread availability of data is the most important shift that has promoted a culture of innovation in the industry.

____

____

2. Industries such as life sciences possess vast scientific and patient data that have gone unused.

____

____

3. It redefines how businesses create value with the help of data.

4. Microsoft pulled the plug on the project only a day after the bot was introduced to the public.

5. The concept of artificial intelligence is making a machine that is no different from you.

II. Translate the following sentences into English.

1. 计算机计算和存储能力的显著提高使人工智能取得了前所未有的突破。

2. 机器学习的发展是惊人的，可以为将来的进一步创新奠定基础。

3. 机器人通过自我学习将拥有回答一切问题的知识。

4. 大数据是一个术语，用来描述公司和政府每天收集的大型数据集。

5. 我们需要使用非传统的技术来收集和处理来自电子商务平台的数据。

Translation at Paragraph Level

English to Chinese Translation

①What 250 years of technological changes have taught us is that technology will have a profound impact on the creation, elimination, and evolution of jobs. ②The advent of AI will reshape jobs in a similar way. ③Twenty-six percent of jobs will be newly created from digital transformation. ④As AI continues to transform the nature of work, we need to rethink education, skills, and training to ensure that people are prepared for the jobs of the future and businesses have access to the talents they need. ⑤And as traditional models of employment transform, we also need to recognize new ways of working and provide adequate protection for workers.

本段一共五句话。第一句是 what 引导的一个主语从句，加上 is 后面由 that 引导的表语从句。on the creation, elimination, and evolution of jobs 是介词短语作状语，翻译时需要前置。第一句可以译为："250 年的技术变革告诉我们，技术对就业岗位的创造、淘汰和演变将产生深远的影响。"第二句是一个简单句，in a similar way 是介词短语作状语，翻译时需要前置。第二句可以译为："人工智能的出现将以类似的方式重塑工作。"第三句是一个被动句，为了保证语句通顺，更符合汉语表达习惯，本句不必译成被动结构。from digital transformation 是介词短语作状语，翻译时需要前置。第三句可以译为："26% 的工作岗位从数字转型中创造出来。"

第四句的主干部分是"We need to rethink education, skills, and training."，句首 as 引导了一个时间状语从句，to ensure 是动词不定式作目的状语，ensure 后面是 that 引导的宾语从句，且宾语从句里用 and 来连接并列结构。for the jobs of the future 是介词短语作状语，翻译时需要前置。they need 是省略了先行词 that 的定语从句，翻译时需要前置。第四句可以译为："随着人工智能继续改变工作性质，我们需要重新思考教育、技能和培训，以确保人们为未来的工作做好准备，企业能够获得所需的人才。"第五句的主句是并列句结构，as 引导了一个时间状语从句。for workers 是介宾结构，翻译时需要前置。第五句可以译为："随着传统就业模式的转变，我们还需要认识新的工作方式，并为工人提供充分的保护。"

III. Translate the following paragraph into Chinese.

While the use of big data will matter across sectors, some sectors are set for greater gains. The computer and electronic products sectors, as well as finance and insurance, and governments will all benefit greatly from the use of big data. Several issues will have to be addressed to capture the full potential of big data. Policies related to privacy, security, and intellectual property will need to be addressed. Organizations need not only the right talents and technologies, but also workflows and incentives to optimize the use of big data.

____

____

____

____

____

____

Chinese to English Translation

①人工智能和机器学习能在新闻机构中扮演什么角色以便人们获得零误差的正确数据？②我们知道在 IP 世界中，数据每 9 至 12 个月就会翻一番，目前没有什么可以阻止这种大规模增长。③它已给全球的企业和服务提供商带来了麻烦。④我们必须阻止那些假新闻，或者为消费者和企业客户提供真实的数据。⑤许多公司都在努力消除假新闻，这是一个亟待解决的问题。⑥过去我们已经看到了太多的假新闻，但现在是时候只展示真实数据了。

本段一共六句话。第一句是一个疑问句，还包含一个目的状语从句“以便……”。“在新闻机构中”作为状语，根据英语表达习惯在翻译时应该后置。第一句可以译为：“What role can artificial intelligence and machine learning play in the news organizations so that people can get correct data with zero error?”。第二句的主干部分是“我们知道……”，可以采用宾语从句的结构，而这个宾语从句里面还需要使用并列结构才能将内容交代清楚。范围状语“在 IP 世界中”和时间状语“目前”根据英语表达习惯在翻译时应该后置。第二句可以译为：“We know that the data doubles every 9 to 12 months in IP world and there is nothing to stop this massive growth at the moment.”。第三句的主干部分是“它已带来了麻烦”，时态为现在完成时。“给全球的企业和服务提供商”可以采用介宾结构作为补语出现。第三句可以译为：“It has created problems for both enterprises and service providers worldwide.”。

第四句是并列句结构，可以使用 or 这个并列连词。“为消费者和企业客户”可以译为介宾结构，根据英语表达习惯在翻译时应该后置。第四句可以译为：“We must stop fake news or provide factual data for consumers and business customers.”。第五句可以把前半句作为主句，时态用现在进行时，后半句则用一个非限定性定语从句来译。第五句可以译为：“Many companies are working hard to eliminate fake news, which is an urgent problem to solve.”。第六句是并列句结构，用连词 but 来表转折。时间状语“过去”根据英语表达习惯在翻译时应该后置。第六句可以译为：“We have seen too much fake news in the past, but it is time to present only real data.”。

IV. Translate the following paragraph into English.

当今世界的数据量呈爆炸式增长，分析大型数据集将成为竞争的关键基础。这将掀起生产率提高、创新和消费者剩余的新浪潮。每个行业的领导者都必须高度重视大数据，而不仅仅是几个数据导向型的管理者。现在数据已经渗透到每一个行业及其业务功能中，成为一个重要的生产要素。领先的公司正在利用数据收集和分析来做出更好的管理决策，其他公司则利用数据及时调整业务。

____

____

____

____

____

____

Workshop

I. Online shopping has developed quickly thanks to the emergence of big data. Make a five-minute oral presentation on how to protect personal information online to the class. An outline is provided for reference.

- Online shopping is a trend.
- The harm of leaking personal information.
- Ways to protect personal information.

II. There's an increasing amount of information available on shoppers and their buying habits. Innovative retailers are looking for ways to tap into big data and give their businesses a competitive advantage. Fill in the table below to find out two useful big data tools for retailers.

No.	Tools	Description	Functions
1.			
2.			

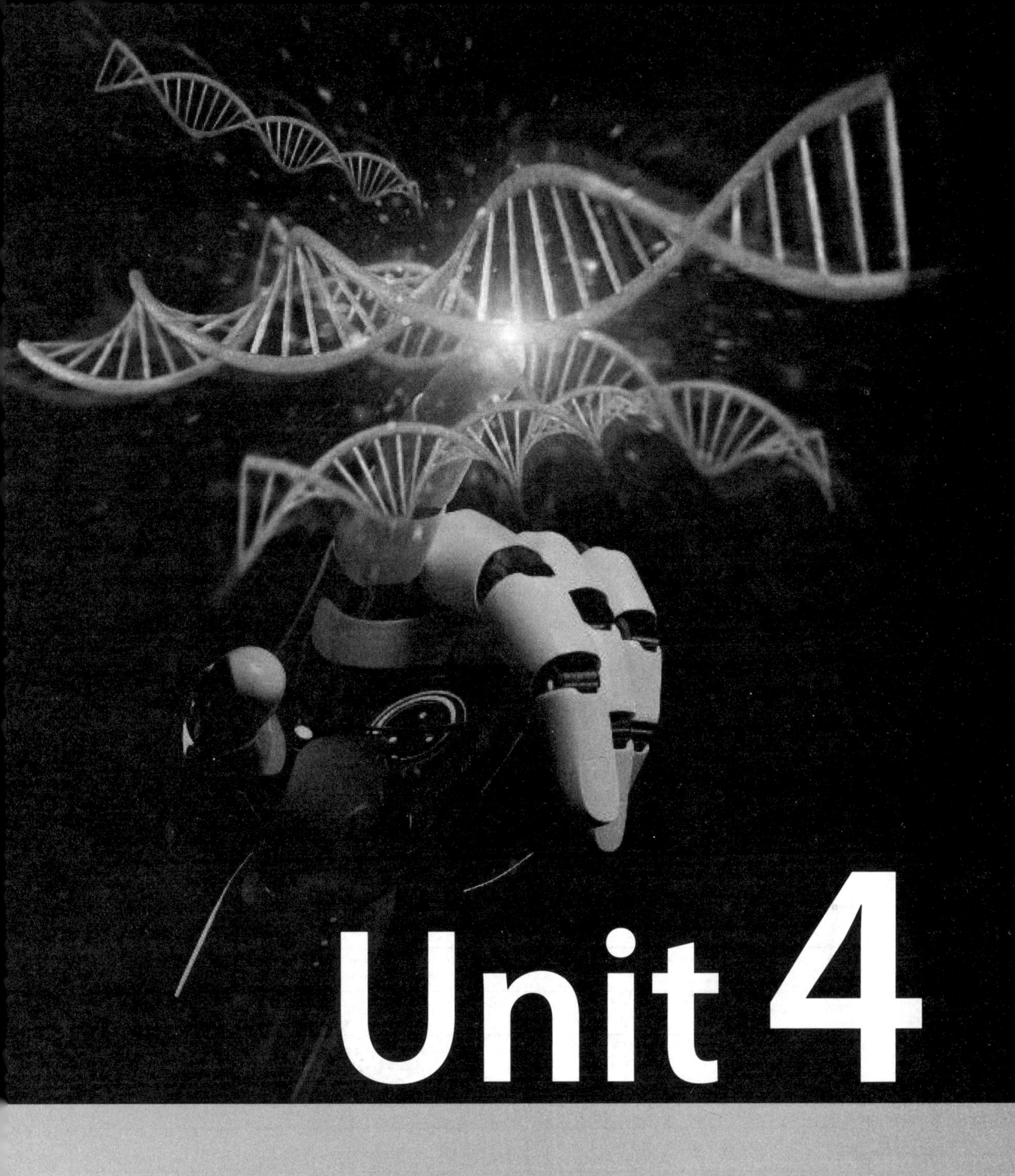

Unit 4

Artificial Intelligence and Cloud Computing

Learning Objectives

In this unit, you will learn:

- an overview of cloud computing;
- the interaction between artificial intelligence and cloud computing;
- key technologies used in the fight against the COVID-19 pandemic;
- writing skills—the usage of the attributive clause;
- translating skills—translation of the attributive clause.

Lead-in

I. Discuss the following questions with your partners.

1. What does "cloud" refer to in the term "cloud computing"?
2. Can you name some famous companies in the field of cloud computing?
3. What is the relationship between artificial intelligence and cloud computing?

II. Work in pairs to discuss the advantages and disadvantages of cloud computing.

Advantages:

1. ____________________
2. ____________________
3. ____________________

Disadvantages:

1. ____________________
2. ____________________
3. ____________________

An Overview of Cloud Computing

1 Cloud computing is a computing **paradigm**, where a large pool of systems are connected in private or public networks, to provide a dynamically **scalable** infrastructure for application, and data and file storage. With the **advent** of this technology, the costs of computation, application hosting, content storage, and delivery are reduced significantly.

2 Cloud computing is a practical approach to directing cost benefits and it has the potential to transform a data center from a capital-intensive setup to a variable-priced environment.

3 The idea of cloud computing is based on a very fundamental principle of "reusability of IT capabilities". The difference that cloud computing brings compared with traditional concepts of "grid computing", "distributed computing", "utility computing", or "autonomic computing" is to broaden horizons across organizational boundaries.

4 Forrester defines cloud computing as "a pool of abstracted, highly scalable, and managed compute infrastructure capable of hosting end-customer applications and billed by consumption".

Models of Cloud Computing

5 Cloud providers offer services that can be grouped into three categories.

Software as a Service (SaaS)

6 In this model, a complete application is offered to the customer, as a service **on demand**. A single instance of the service runs on the cloud and **multiple** end users are serviced. On the customer's side, there is no need for **upfront** investment in servers or software licenses, while for the provider, the costs are lower, since only a single application needs to be hosted and maintained. Today, SaaS is offered by companies such as Google, Salesforce, Microsoft, Zoho, etc.

Platform as a Service (PaaS)

7 Here, a layer of software or development environment is **encapsulated** and offered as a service, upon which other higher levels of services can be built. The customer has the freedom to build his own applications, which run on the provider's infrastructure. To meet manageability and scalability requirements of the applications, PaaS providers offer a predefined combination of OS and application servers, such as LAMP platform (Linux, Apache, MySQL, and PHP), restricted

J2EE, Ruby, etc. Google's App Engine, Force.com, etc. are some of the popular PaaS examples.

Infrastructure as a Service (IaaS)

8 IaaS provides basic storage and computing capabilities as standardized services over the network. Servers, storage systems, networking equipment, data center space, etc. are pooled and made available to handle workloads. The customer would typically **deploy** his own software on the infrastructure. Some common examples are Amazon, GoGrid, 3 Tera, etc.

Understanding Public and Private Clouds

9 Enterprises can choose to deploy applications on public, private, or hybrid clouds. Cloud integrators can play a vital part in determining the right cloud path for each organization.

Public Cloud

10 Public clouds are owned and operated by third parties; they deliver superior economies of scale to customers, as the infrastructure costs are spread among a mix of users, giving each individual client an attractive low-cost, "pay-as-you-go" model. All customers share the same infrastructure pool with limited **configuration**, security protections, and availability variances. These are managed and supported by the cloud provider. One of the advantages of a public cloud is that it may be larger than an enterprise's cloud, thus providing the ability to scale **seamlessly**, on demand.

Private Cloud

11 Private clouds are built exclusively for a single enterprise. They aim to address concerns on data security and offer greater control, which is typically lacking in public clouds. There are two variations of private clouds:

- On-premise private clouds, also known as internal clouds, are hosted within one's own data center. This model provides a more standardized process and protection, but is limited in terms of size and scalability. IT departments would also need to incur the capital and operational costs for the physical resources. This is best suited for applications which require complete control and configurability of the infrastructure and security.
- Externally hosted private clouds are hosted externally with a cloud provider, where the provider facilitates an exclusive cloud environment with full guarantee of privacy. This is best suited for enterprises that don't prefer a public cloud due to sharing of physical resources.

Hybrid Cloud

12 Hybrid clouds combine both public and private cloud models. With a hybrid cloud, service providers can utilize the third party cloud providers in a full or partial manner, thus increasing the flexibility of computing. The hybrid cloud environment is capable of providing on-demand, externally provisioned scale. The ability to **augment** a private cloud with the resources of a public cloud can be used to manage any unexpected surges in workload.

Benefits of Cloud Computing

13 Enterprises would need to **align** their applications, so as to exploit the architecture models that cloud computing offers. Some of the typical benefits are listed below.

Reduced Cost

14 There are many reasons to attribute cloud technology to lower costs. The billing model is pay-as-per-usage; the infrastructure is not purchased, thus lowering maintenance cost. Initial expense and recurring expense are much lower than traditional computing.

Increased Storage

15 With the massive infrastructure that is offered by cloud providers today, storage and maintenance of large volumes of data is a reality. Sudden workload **spikes** are also managed effectively and efficiently, since the cloud can scale dynamically.

Flexibility

16 With enterprises having to adapt, even more rapidly, to changing business conditions, speed to deliver is critical. Cloud computing stresses on getting applications to market very quickly by using the most appropriate building blocks necessary for deployment.

Challenges of Cloud Computing

17 Despite its growing influence, concerns regarding cloud computing still remain. In our opinion, the benefits outweigh the drawbacks and the model is worth exploring. Some common challenges are listed below.

Data Protection

18 Data security is a crucial element that warrants **scrutiny**. Enterprises are reluctant to buy an assurance of business data security from vendors. They fear losing data to competitors and the data **confidentiality** of consumers. In many instances, the actual storage location is not disclosed, adding to the security concerns of enterprises. In the cloud model, service providers are responsible for maintaining data security and enterprises would have to rely on them.

Data Recovery and Availability

19 All business applications have service-level agreements that are stringently followed. Operational teams play a key role in management of service-level agreements and runtime governance of applications. In production environments, operational teams support:

- appropriate clustering and failover;
- data replication;
- system monitoring (transactions monitoring, logs monitoring, and others);
- maintenance (runtime governance);
- disaster recovery;

- capacity and performance management.

20 If any of the above-mentioned services is under-served by a cloud provider, the damage and impact could be severe.

Management Capabilities

21 Despite there being multiple cloud providers, the management of platform and infrastructure is still **in its infancy**. There is huge potential to improve the scalability and load balancing features provided today.

Regulatory and Compliance Restrictions

22 In some European countries, government regulations do not allow customers' personal information and other sensitive information to be physically located outside the state or country. To meet such requirements, cloud providers need to set up a data center or a storage site exclusively within the country to **comply with** regulations. Having such an infrastructure may not always be feasible and is a big challenge for cloud providers.

23 With cloud computing, the action moves to the interface between service suppliers and multiple groups of service consumers. Cloud services will demand expertise in distributed services, **procurement**, risk assessment, and service negotiation—areas that many enterprises are only modestly equipped to handle.

Notes

Forrester an American market research company that provides advice on existing and potential impact of technology for its clients and the public.

Salesforce an American cloud-based software company headquartered in San Francisco, California. Though the bulk of its revenue comes from a customer relationship management product, Salesforce also sells a complementary suite of enterprise applications focusing on customer service, marketing automation, analytics, and application development.

Zoho an Indian software development company. The focus of Zoho Corporation lies in Web-based business tools and information technology solutions, including an office tools suite, Internet of Things (IoT) management platform, and a suite of IT management software.

LAMP platform an archetypal model of Web service stacks, named as an acronym of the names of its original four open-source components: the Linux operating system, the Apache HTTP Server, the MySQL relational database management system (RDBMS), and the PHP programming language. The LAMP components are largely interchangeable and not limited to the original selection. As a solution stack, LAMP is suitable for building dynamic websites and Web applications.

Linux an operating system that evolved from a kernel created by Linus Torvalds when he was a student at the University of Helsinki. To say that Linux is an operating system means that it's meant to be used as an alternative to other operating systems, Windows, Mac OS, MS-DOS, Solaris, and others. Linux is not a program like a word processor and is not a set of programs like an office suite. Linux is an interface between computer and server hardware, and the programs which run on it.

Apache known as Apache HTTP Server, an open-source Web server provided by Apache Software Foundation that can be run on most operating systems.

MySQL an open-source relational database management system (RDBMS), which is the world's most popular open-source RDBMS and is currently ranked as the second most popular RDBMS in the world (behind Oracle Database).

PHP known as Pre Hypertext Preprocessor, an open-source, server-side, HTML embedded scripting language used to create dynamic Web pages.

J2EE known as Java 2 Platform Enterprise Edition, a platform-independent, Java-centric environment from Sun for developing, building, and deploying Web-based enterprise applications online. The J2EE platform consists of a set of services, APIs, and protocols that provide the functionality for developing multi-tiered, Web-based applications.

Ruby a simple but powerful object-oriented programming language. Ruby programs are compact, yet readable and maintainable. Ruby offers features such as blocks, iterators, meta-classes, and others, and it can be used to write servers, to experiment with prototypes, and for everyday programming tasks. Ruby is open-source and freely available for both development and deployment.

Google's App Engine a Web framework and cloud computing platform for developing and hosting Web applications in Google-managed data centers. Applications are sandboxed and run across multiple servers. App Engine offers automatic scaling for Web applications—as the number of requests increases for an application, App Engine automatically allocates more resources for the Web application to handle the additional demand.

Force.com the website of Salesforce mentioned above.

Amazon a Seattle, Washington-based, multinational technology company focusing on e-commerce, cloud computing, and artificial intelligence.

GoGrid a cloud infrastructure service, hosting Linux and Windows virtual machines managed by a multi-server control panel and a RESTful API.

3 Tera a developer of system software for utility computing and cloud computing. It is headquartered in Aliso Viejo, California. It was acquired by CA Technologies in 2010.

Words and Expressions

paradigm	/ˈpærədaɪm/	*n.*	a typical example or pattern of something 范式，样例
scalable	/ˈskeɪləbl/	*adj.*	able to be used or produced in a range of capabilities 可用于（或产生）多种能力的
advent	/ˈædvent/	*n.*	the approach or arrival of (an important person, event, etc.) 出现，到来
on demand			according to needs 按需
multiple	/ˈmʌltɪpl/	*adj.*	having or involving many individuals, items, or types 许多的，多样的；多重的
upfront	/ˌʌpˈfrʌnt/	*adj.*	paid before any work has been done or before goods are supplied 预付的
encapsulate	/ɪnˈkæpsjuleɪt/	*v.*	to put in a short or concise form 压缩
deploy	/dɪˈplɔɪ/	*v.*	to use something effectively 有效地利用；调动
configuration	/kənˌfɪgəˈreɪʃn/	*n.*	the combination of equipment needed to run a computer system 配置
seamlessly	/ˈsiːmləsli/	*adv.*	done or made so smoothly that you cannot tell where one thing stops and another begins 流畅地，不间断地
augment	/ɔːgˈment/	*v.*	to increase the value, amount, effectiveness, etc. of something 加强；提高；增大
align	/əˈlaɪn/	*v.*	to change something slightly so that it is in the correct relationship to something else 使一致
spike	/spaɪk/	*n.*	a sudden large increase in the number or rate of something 激增
scrutiny	/ˈskruːtəni/	*n.*	careful and thorough examination of someone or something 详细审查；监视；细看
confidentiality	/ˌkɒnfɪˌdenʃiˈæləti/	*n.*	a situation in which you trust someone not to tell secret or private information to anyone else 机密；保密性
in one's infancy			at the preliminary stage 处于初期
regulatory	/ˈregjələtəri/	*adj.*	restricting according to rules or principles 监督的；

监管的

comply with to abidy by 遵守

procurement /prə'kjʊəmənt/ *n.* the act of getting possession of something 采购；获得

Useful Terms

grid computing	网格计算
distributed computing	分布式计算
utility computing	效用计算
autonomic computing	自主计算
Software as a Service (SaaS)	软件即服务
Platform as a Service (PaaS)	平台即服务
Infrastructure as a Service (IaaS)	基础设施即服务
pay-as-you-go	现收现付，即付即用

Exercises

Comprehension Check

I. Answer the following questions according to the text.

1. What is the basis of the cloud computing concept?
2. What are the major models of cloud computing?
3. What are the differences between public clouds and private clouds?
4. What are the benefits of cloud computing?
5. What are the challenges of cloud computing?

II. Read the following statements carefully and decide whether they are true (T) or false (F) without turning back to check the text.

1. ______________ Cloud computing is believed to be effective in decreasing the costs in computation and applications of IT companies.

2. ______________ In the SaaS cloud computing model, the customer can freely create his application and share with other end users.

3. ______________ Public clouds are financed, owned, and managed by the government to provide services.

4. ______________ In the cloud computing models, there is no need to worry about privacy and information security.

5. ______________ Cloud computing is worth exploring with more benefits than drawbacks.

III. Choose the best answer to each of the following questions according to the text.

1. Cloud computing can be applied in the following fields EXCEPT ______________.

   A. complex engineering computation

   B. hosting end-customer applications

   C. content storage and delivery

   D. free access to personal information

2. Compared with traditional computing concepts, cloud computing is particular in ______________.

   A. bringing differences

   B. distributing information to customers

   C. conducting computation autonomously

   D. breaking down organizational boundaries and broadening horizons

3. The role of determining the right cloud path for the organization is largely played by ______________.

   A. the cloud integrator

   B. the public cloud provider

   C. the private cloud provider

   D. the hybrid cloud provider

4. Which of the following statements is a factor contributing to the enterprises' security concerns?

   A. There is no firewall on the cloud to protect sensitive information.

   B. The enterprises do not know where their information is stored.

   C. The government is not involved.

D. The enterprises are not willing to pay for the insurance.

5. Who is responsible for the management of service-level agreements?

A. The board of directors.

B. The R&D department.

C. The operational teams.

D. The sales teams.

Vocabulary Building

IV. Fill in the following blanks with the words and phrases given in the box. Change the form if necessary.

advent	align	deploy	configuration	augment
scrutiny	regulatory	on demand	in one's infancy	comply with

1. Once you compile the code, you need to ______________ it in the server to access the service.

2. The evaluation process should also take into consideration experience gained by national ______________ authorities in clearing such products.

3. In fact, with the ______________ of paid inclusion, some search engines now have a vested interest in the health of the optimization community.

4. That's why we are open sourcing it at an early stage: We believe that peer review, community feedback, and public ______________ greatly improve the quality of security technologies like this one.

5. These are preinstalled so that you can test and use them without spending time on installation and ______________.

6. The quick evolution of the business makes it impossible for IT to ______________ with it currently.

7. If we wanted to ______________ the set of commands, we would need to make some code changes.

8. If we reinforce the traffic security education and more people ______________ traffic regulations, I believe one day traffic accidents can be avoided.

9. These services can be provided by an intermediary through which interested businesses can route their messages ______________.

10. To be fair, escape analysis is still ______________ and it will take time to mature this complex tool.

V. Match the words in the left column with the explanations in the right column.

1.	potential	A.	sudden occurrence or increase in amount or number
2.	exclusive	B.	to use or develop something fully and effectively
3.	surge	C.	possibility of being developed or used
4.	exploit	D.	special knowledge or skill, especially in a particular field
5.	expertise	E.	only available or belonging to particular people, and not shared

Word Formation

VI. Fill in the following blanks with the words in capitals. Change the form if necessary. An example has been given.

e.g.	*All languages must have some capability to control the flow of* *computation and represent data.*	**COMPUTE**
1.	We received training on a number of spreadsheet and database ______________.	**APPLY**
2.	There is a need for ______________ and distribution of the meeting materials among different departments.	**REPLICATE**
3.	The law imposed new financial ______________ on private companies.	**RESTRICT**
4.	If this is the initial ______________, the server should not be running.	**CONFIGURE**
5.	An agreement was reached between the two companies after several rounds of lengthy ______________.	**NEGOTIATE**

Translation

VII. Translate the following sentences into English with the words and phrases in brackets.

1. 随着云计算技术的出现，计算、应用托管、内容存储和传递的成本能显著降低。(advent; delivery; significantly)

____

____

2. 云集成商在为每个机构确定正确的云路径方面发挥着至关重要的作用。(vital; determine)

____

____

3. 企业需要调整他们的应用程序，以便利用云计算提供的架构模型。(align; exploit)

____

____

Text B

How Artificial Intelligence Is Transforming Cloud Computing

Cloud computing has already ***permeated*** *every facet of online activity. However, recent developments in AI and the increasing sophistication of programmers* ***presage*** *a new age of cloud computing. This article looks at how that technology emerges and how it's set to impact our lives.*

1 Every tech **guru** knows about the potential of cloud technology and how it has already affected the ways businesses and citizens store data and existing workloads. But because the cloud is such a new technology, companies have to think about how it will continue to evolve over time. Trends like the rise of mobile in place of computers and the Internet of Things have made small changes to cloud technology. But now the big dream is how artificial intelligence could improve cloud technology just as cloud technology has improved AI development.

2 IBM, one of the biggest cloud companies out there, states that the **fusion** of AI and cloud computing "promises to be both a source of innovation and a means to accelerate change". The cloud can help provide AI with the information which it needs to learn, while AI can provide information which can give the cloud more data. This **symbiotic** relationship can transform our development of AI, and the efforts of cloud companies like IBM to **delve into** AI research show that these are not empty words.

Can AI Finally Learn?

3 One of the biggest transformations in AI development has been how technology companies can create AI which can finally learn. One well-known example of this new machine learning occurred earlier this year when an AI defeated the world's best Go player. Instead of **brute-forcing** the best moves like DeepBlue did in chess about 20 years ago, the AI learned by playing millions of games with itself and figuring out strategies which even Go players had not considered.

4 Of course, machine learning has far more useful practical purposes than playing games. One of the biggest fields is in conversation, where voice-responsive AI systems can respond to human commands.

5 While we already have personal assistants like Cortana which can respond to voice commands, technology companies are interested in developing AI systems which can learn new words and how to respond differently. While there is still a lot of work to be done as the disastrous Tay experiment from earlier this year shows, the goal is to construct an AI which can communicate like a normal human.

6 Cloud computing could help immensely with this goal. Many **disparate** servers which are part of cloud technology hold the data which an AI can access and use to make decisions and learn things like how to hold a conversation. But as the AI learns this, it can **impart** this new data back to the cloud, which can thus help other AIs learn as well.

Combining AI and the Cloud

7 As noted above, the potentials of AI and the cloud mean that companies which **specialize in** one of those two are putting more work into getting involved with both. As *CIO* notes, many of these cloud AI technologies **take on** two forms. They are cloud machine learning platforms like Google Cloud Machine Learning which combine machine learning with the cloud but do not have deep learning frameworks, or they are AI cloud services like IBM Watson.

8 The latter in particular is interesting because of the numerous applications through which businesses can use AI cloud services. For example, *Wired* recently reported about how organizations are relying on IBM Watson to help fight cybercrime. This is not as simple as plugging Watson into a USB and letting it go to work. Researchers have to teach Watson various parts of how to deal with hackers and criminals, where it becomes steadily more effective as it stores information through the network.

9 What is incredible about this learning process is that while Watson knows so much, there is still an important role for humans to play. Watson could read far more reports than a human could, but he still makes basic, fundamental mistakes like thinking that "ransomware" indicates a place. The researchers help Watson and guide the data so that he learns to think correctly.

10 In every step of the way, AI, cloud technology, and humans are all needed. AI is needed to

learn, cloud technology is needed so that AI can access more data about cybercrime than what could be stored on a server, and humans help AI when it makes mistakes. This sort of cooperation and technological development can apply to almost any field which we can think of today, as well as some which we may not even **conceive of** now.

11 Combining AI, machine learning, and the data stored with cloud technology means that both AI and humans can analyze and gather more data than ever before. Tech experts have indicated that 2017 could be the year when AI becomes a **ubiquitous** part of our daily lives, and AI capabilities will only be improved with the development of cloud technology.

12 So pay attention to companies like Google and IBM as they work on combining the two. The result will be a world which transforms how we view both AI and the cloud.

Notes

Cortana a virtual assistant created by Microsoft for Windows 10, Windows 10 Mobile, Windows Phone 8.1, Invoke Smart Speaker, Microsoft Band, Surface Headphones, Xbox One, iOS, Android, Windows Mixed Reality, and Amazon Alexa. Cortana can set reminders, recognize natural voice without the requirement for keyboard input, and answer questions using information from the Bing search engine.

Tay an artificial intelligence chatbot that was originally released by Microsoft via Twitter on March 23, 2016. It caused subsequent controversy when the bot began to post inflammatory and offensive tweets through its Twitter account, forcing Microsoft to shut down the service only 16 hours after its launch. According to Microsoft, this was caused by trolls who "attacked" the service as the bot made replies based on its interactions with people on Twitter. It was soon replaced with Zo.

CIO a magazine founded in 1987, which is now entirely digital (www.CIO.com). CIO refers to the job title: chief information officer. Readership includes other technology-related executives and IT decision makers.

Wired a monthly American magazine, published in print and online editions, which focuses on how emerging technologies affect culture, economy, and politics.

Words and Expressions

permeate /ˈpɜːmieɪt/ *v.* to enter something and spread through every part of it 渗透，透过；弥漫

presage	/ˈpresɪdʒ/	*v.*	to be a sign that something is going to happen, especially something bad 预示，预兆；预感
guru	/ˈɡʊruː/	*n.*	someone who knows a lot about a particular subject, and gives advice to other people 专家，权威
fusion	/ˈfjuːʒn/	*n.*	a combination of separate things, qualities, or ideas 融合，合成
symbiotic	/ˌsɪmbaɪˈɒtɪk/	*adj.*	a relationship between two different living creatures that exist together in a way of depending on each other 共生的
delve into			to study intensively 探索，钻研
brute-force	/bruːt ˈfɔːs/	*v.*	to make somebody do something he/she does not want to do 强迫，逼迫，迫使
disparate	/ˈdɪspərət/	*adj.*	different and not related to each other 迥然不同的；无法比较的
impart	/ɪmˈpɑːt/	*v.*	to pass information, knowledge, etc. to other people 透露；传授
specialize in			to limit all or most of your study, business, etc. to a particular subject or activity 专门从事，专攻
take on			to begin to have a particular quality or appearance 呈现
conceive of			to think of an idea or a plan and work out how it can be done 想象；设想
ubiquitous	/juːˈbɪkwɪtəs/	*adj.*	seeming to be present everywhere or in several places at the same time 普遍存在的，无处不有的

Useful Terms

USB	通用串行总线
ransomware	勒索软件

Exercises

Comprehension Check

I. Identify the paragraph from which the information is derived.

1. ______________ Companies need to consider the trend of the development of cloud technology since it is new.
2. ______________ Technology companies show enthusiasm in developing AI systems that can learn new words and give different responses.
3. ______________ AI cloud services are especially interesting to businesses in that they can use such services through many applications.
4. ______________ There are mainly two forms of cloud AI technologies according to *CIO*.
5. ______________ It has been indicated that the year of 2017 could witness the wide spreading of AI in people's daily life.
6. ______________ According to IBM, the blending of AI and cloud computing is expected to bring innovation and speed up changes.
7. ______________ Throughout the AI learning process, AI, cloud technology, and humans are all needed.
8. ______________ A big transformation in AI development has been that technology companies are able to create self-learning AI systems.
9. ______________ Machine learning has a very useful practical application in conversation.
10. ______________ Google and IBM could be the leaders in the fusion of AI and cloud technology.

II. Answer the following questions according to the text.

1. How does IBM explain the relationship between AI and cloud computing?
2. What is the goal of present machine learning application in conversation?
3. How can cloud computing help construct an AI which can communicate like a normal human?
4. What are the major forms of cloud AI technologies according to *CIO*?
5. What is the relationship among AI, cloud technology, and humans in the process of combining AI and the cloud?

Vocabulary Building

III. Fill in the following blanks with the words and phrases given in the box. Change the form if necessary.

permeate	fusion	disparate	impart	accelerate
delve into	specialize in	respond to	take on	conceive of

1. For that reason, it was only a matter of time until the technologies of the network would reinforce this natural tendency and ______________ it.
2. A component that calls the service interface can ______________ all types of data.
3. Shortly after suffering from a massive earthquake, the city ______________ a new look.
4. The software teams had great talent and a clear road map, and Steve Jobs' attention to detail and passion for perfection ______________ Apple's culture from top to bottom.
5. She managed to ______________ great elegance to the plain dress she was wearing.
6. A team may be made up solely of analysts who ______________ requirements gathering and use case development.
7. He adds that tomorrow's world will be a(n) ______________ of biology and technology, where robots do the chores, cars drive themselves, and artificial limbs are better than real ones.
8. In 1900, nobody could ______________ digital electronics, computers, and their infinite variations; these inventions were literally unimaginable.
9. There will be instances where you simply must relate the data from ______________ data sources for certain reports.
10. Although these articles build on each other as a series, each stands on its own if you want to ______________ a particular area of interest.

Text C

Technologies Critical in the Fight Against COVID-19 Pandemic

1 The world has been severely hit by the pandemic triggered by the widespread novel coronavirus. **All walks of life** have been forced to **overhaul** their lifestyle and work habits. Such transformation couldn't have been possible without cloud-based services, which allow enterprises to swiftly alter their business models to catch up with these rapid changes.

> All walks of life 各行各业
> overhaul *v.* 全面改革

2 Imagine if there were no cloud, the "**lockdown**" would likely keep businesses out of operation. Working from home would be one of the wildest dreams. Online streaming entertainment like Netflix would not be helping so many **grounded** people with entertainment. Not to mention delivery services like Deliveroo that bring food to families that normally don't spend much time in the kitchen. All these require robust AI and cloud services to keep things running as "normally" as we can hope for.

> lockdown *n.* 防范禁闭；封锁
> grounded *adj.* 禁足的

Cloud Helps People to Live Normally amid Abnormal Times

3 But cloud-based services do more than just let people live as normally as possible in such unprecedented times. The power of cloud has been shown in a wide range of scenarios. It can be as simple as online meetings, or some complex tasks requiring AI and cloud computing, such as **gene sequencing**, drug screening, diagnosis, and epidemic survey. It can be used to monitor whether people are properly wearing masks. Everything counts.

> gene sequencing 基因测序

4 In terms of education, with so many students grounded, online learning becomes essential, and it requires solid cloud-based services. It is more than just providing cloud space for virtual meetings; it also requires **a host of** tools to maintain teaching standards.

> a host of 许多，大量

5 As an international **ICT** enterprise, Huawei is devoted to making the world better. Among those measures is the **TECH4ALL** corporate program, which aims to help everyone gain access to technology. Amid this pandemic, we see the urgent need of technology to ensure that **deliverables** reach the right places.

abbr. 信息通信技术
华为“数字包容”计划
n. 可交付成果

Various Sectors Can Benefit from Cloud-Based Services

6 Among different measures under the international action plan, Huawei Cloud is now providing AI and cloud services to help customers around the world fight epidemics. The initiatives cover various domains ranging from medical and educational to enterprise supports. For education, Huawei Cloud is currently working with various partners to provide online education services for schools and other educational institutions. More than 1,000 schools are using the Huawei Cloud online education solution, and the results have been impressive.

7 For healthcare, Huawei Cloud is providing **EIHealth** services. Powered by the advantages of AI and big data technologies, EIHealth provides a professional AI R&D platform to accelerate AI applications in **genomics**, drug discovery, and medical imaging. The free services include **viral genome detection**, anti-viral drug in silico screening, and AI-assisted **CT** patient screening service, and more. Since January, the Huawei Cloud AI solution has been used at more than 100 medical institutions.

医疗智能体
n. 基因组学；基因体学
病毒基因组检测

8 In the business sector, enterprises are also receiving support to **migrate** businesses to the cloud to ensure continuous operations while the pandemic continues. Huawei Cloud is now offering a 12-month package with up to 1,500 hours of free cloud resources for each newly registered user, **complete with** 24/7 professional support online. In February, Huawei Cloud launched a special program, providing free cloud resource packages for **small and medium-sized enterprises** (SMEs) during the pandemic. Other solutions are also ready to be in place to support **work resumption**.

v. 移动，转移
包括
中小企业
复工

Joint Efforts Necessary to Fight Pandemic

9 Because it will take collective efforts to **weather** this hardship, Huawei is hoping more organizations can get together to help everyone in the critical moments. In fact, significant progress has been seen

v. 经受住

in many regions, where Huawei is working with partners to provide support for hospitals, medical care facilities, and SMEs.

10 In Malaysia, for instance, Huawei works with the country's Ministry of Health to contribute the Huawei Cloud AI-assisted diagnosis solution to Sungai Buloh Hospital. The goal is to empower local medical personnel with AI capabilities by providing an AI solution for CT image analysis of possible COVID-19 patients. ULearning of Indonesia is providing online education in local universities via Huawei's solutions, while Singapore's 7-Network provides a health information report platform for SMEs with support from Huawei Cloud.

11 Other countries including Ecuador and Panama are using AI technology to screen COVID-19 in several hospitals, while Argentina International Airport is applying an intelligent **thermal imaging temperature measurement system**. These technologies are launched by Huawei Cloud and local partners.

热成像测温系统

12 At the same time, Huawei is also supporting numerous initiatives led by other private organizations to fight COVID-19 by providing cloud and AI technologies as well as tech support.

13 Strongly believing in partnership, Huawei hopes more organizations can work together to not only help fight the epidemic, but also help the world recover within a reasonable time frame. Only unity will help us win this war.

I. Answer the following questions according to the text.

1. What is the impact of the novel coronavirus pandemic?

2. What can cloud-based services do to help people live normally during the pandemic?

3. What is the purpose of Huawei's TECH4ALL program?

4. What services can Huawei Cloud provide in the healthcare field?

5. What does Huawei think is necessary to win the war against COVID-19?

II. Read the following statements carefully and decide whether they are true (T) or false (F) without turning back to check the text.

1. ______________ Cloud technology contributed a lot to coping with the influence of COVID-19.

2. ______________ Cloud-based learning services refer to providing cloud space for teachers and students to have virtual classes.

3. ______________ The application of Huawei Cloud in online education has proved effective.

4. ______________ Huawei Cloud services are only available in Asian countries.

5. ______________ Huawei expects to cooperate with partners to fight COVID-19.

III. Discuss the following questions based on your understanding of cloud computing.

1. Have you ever used any Huawei Cloud service? If yes, please share your experience with your partners. If not, please search some information and share it with your partners.

2. Did you use any cloud-based learning service during the pandemic? Please share your experience with your partners.

3. What other cloud-based services can make your life and study more convenient and efficient?

Writing Skills

定语从句

与普通英语相比，科技英语的语法关系显得更为复杂，其原因在于科技英语中有许多从句，尤其是定语从句（attributive clause）。

定语从句又称关系从句（relative clause），用于修饰一个名词、代词或名词性短语，有时也可以修饰一个句子。被定语从句修饰的名词、代词或短语叫先行词，定语从句通常跟在先行词的后面。

定语从句通常由关系代词 that、which、who、whom、whose 和关系副词 when、where、why、how 引导。关系代词和关系副词通常放在先行词和定语从句之间，起连接作用，同时还代替先行词在句中充当一定的语法成分，如主语、宾语、定语和状语等。

例 1：Computer programs that can be run by a computer's operating system are called executables.

在例 1 中，that can be run by a computer's operating system 是定语从句，computer programs 是先行词。

基于先行词限制意义的强弱，定语从句可以分为限定性和非限定性两种。

限定性定语从句与先行词关系密切，是整个句子中不可缺少的部分。如果少了它，句子的意思就不完整或不明确。这种定语从句与主句之间不用逗号隔开。当限定性定语从句修饰人时，一般用关系代词 who，有时也用 that。若关系代词在定语从句中作主语，则 who 用得较多，且不可省略。若关系代词在定语从句中作宾语，则应当使用宾格 whom 或 that，但在大多数情况下都可以省略。若关系代词表示所属关系，则应用 whose。当限定性定语从句修饰物时，一般用 that 较多，也可以用 which。它们可以在从句中作主语，也可以作宾语；若作宾语，则大多数情况下可以省略。

例 2：Mouse is an instrument (which) operators often use.

在例 2 中，which 引导的定语从句修饰 an instrument。因为 which 在从句中作 use 的宾语，故可省略。

例 3：PCTOOLS are tools whose functions are very advanced.

在例 3 中，因为 functions 和 tools 之间是所属关系，故用所有格 whose。

非限定性定语从句与先行词的关系比较松散，从句只是对先行词进行的附加说明。如果缺少它，并不会影响整个句子的主要意思。这种定语从句与主句之间常用逗号隔开。非限定性定语从句在修饰人时用 who、whom 或 whose 引导，修饰物时用 which 引导，修饰句子时用 which 或 as 引导，修饰地点和时间时分别用 where 和 when 引导。关系代词 that 和关系副词 why 不能引导非限定性定语从句。

例 4：We do experiments with a computer, which helps to do many things.

在例 4 中，which 引导的非限定性定语从句是对先行词 a computer 的说明。

Change the following sentences into compound sentences with attributive clauses.

1. This is a College of Science and Technology. The students of this college are trained to be engineers or scientists.

   ____

   ____.

2. Electronic computers have many advantages. But they cannot carry out creative work and replace man.

____

____.

3. A computer is an electronic device. This device can receive a set of instructions, or a program, and then carry out this program by performing calculations on numerical data or by compiling and correlating other forms of information.

____

____.

4. Systems such as eCash allow the customer to deposit cash into a bank account and receive an encoded digital string for each monetary unit. This digital string is then transferred into the customer's hard disk.

____

____.

5. To participate in e-mails, a person must be assigned a mailbox. The mailbox is in fact a storage area. Messages can be placed in this area.

____

____.

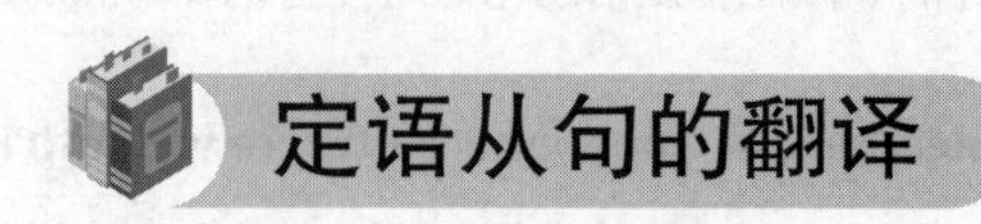

英语重形合，为了对一个名词或名词词组进行完整、明确的阐述，通常借助各种修饰结构（介词短语、不定式短语、分词短语、同位语或定语从句等）引出一长串修饰成分。汉语重意合，各个成分用意义串联，通常不需要连接词，句子一般简短明快，很少出现长句。对于一个名词，汉语通常不使用叠床架屋式的多重修饰语，因为前置定语过长会使句子变得冗长、臃肿，甚至晦涩难懂。

由于英汉两种语言在修饰方式上的差异，对于科技英语中含有多重定语或长定语从句的中心词（组），在翻译时应采取拆零、化整为零、破句重组等方法，避免使用过长的修饰语。具体可以参考以下四种翻译方法：

1. 前置法

定语从句在语法结构上是作定语修饰语，所以在翻译时，通常把较短的定语从句译成带“的”的前置定语，放在定语从句的先行词前面，从而将英语复合句译成汉语单句。

例 1：In the room where the electronic computer is kept, there must be no dust at all.
存放电子计算机的房间里，不能有一点灰尘。

2. 后置法

当定语从句的结构较长且复杂时，如果采用前置法翻译会造成定语过于臃肿，且叙述不清。这时，为使译文连贯、流畅，我们可以采用主句和从句分开翻译的方法，即把定语从句译成并列分句，放在先行词的后面。翻译时可以采用两种方法来处理：

第一，重复先行词。由于先行词通常在定语从句中充当句子的成分，在单独翻译定语从句时通常需要重复先行词，并可以用“这”“那”“其”“它”等作为独立句的主语。

例 2：Between these two tiny particles, the proton and the electron, there is a powerful attraction that is always present between negative and positive electric charges.
在质子和电子这两个微粒之间有一个很大的吸引力，而这个吸引力总是存在于正、负电荷之间。

例 3：A sound card is a printed circuit board that can translate digital information into sound and back, that plugs into a slot on the motherboard (the main circuit board of a computer), and that is usually connected to a pair of speakers.
声卡是一块印刷电路板，它能把数字信息译为声音，也能把声音变为数字信息；它插在母板（计算机主电路板）上的槽内，且通常连有一对喇叭。

第二，省略先行词。当把定语从句放在先行词的后面翻译时，在保证句子通顺、完整的前提下，有时候可以不用重复先行词。

例 4：They worked out a new method by which production has now been rapidly increased.
他们制订了一种新方案，采用之后产量已迅速提高。

3. 融合法

融合法是指翻译时打破原文的句子结构，译者根据对原文的理解用自己的话译出内容。除了通常用于 there be 句型之外，这种方法还适用于另外一种情况，即英语主句比较简单，重点在于定语从句。这时，译者可以把主句译成简单句的主语，把定语从句译成谓语。由于限定性定语从句与主句关系较紧密，融合法多用于翻译这类定语从句。

例 5：There are some metals which possess the power to conduct electricity and the ability to be magnetized.
某些金属既能导电，又能被磁化。

例 6：We used a computer of which almost every part carried some identification of national identity.
我们使用的电脑几乎每一个部件都有国籍的某个标志。

4. 分译法

分译法是指把主句和从句分开翻译的一种方法，主要用于句子结构较长的非限定性定语从句。由于非限定性定语从句与先行词之间的关系较为疏远，我们通常会把定语从句译成独立的句子，以避免句子过于冗长和累赘。

例 7：The users of such a system control the process by means of a program, which has a set of instructions that specify the operation, operands, and the sequence by which processing has to occur.
该系统用户通过程序控制处理过程，所谓程序是一套指定操作、操作数和处理序列的指令集。

实际上，英语也存在意合情况。在科技英语中，有些从句虽然从语法结构上看是定语从句，但实际上起着修饰各种逻辑状语的作用，如表示原因、结果、让步、假设等。这种情况在科技英语中很常见，翻译时稍有不慎则易引起歧义。在翻译这类句子时，可以从定语从句的逻辑性质和功能作用出发，参考以下翻译方法，适当地译成汉语的偏正复句。

例 8：You must grasp the concept of "work", which is very important in physics.（表原因）
你必须掌握"功"的概念，因为它在物理学中很重要。

例 9：Copper, which is used so widely for carrying electricity, offers very little resistance.（表结果）
铜的电阻很小，所以被广泛地用来传输电力。

例 10：He insisted on designing another application, which we had no use for.（表让步）
他坚持要再设计一个应用，尽管我们并无此需要。

例 11：For any machine, whose input and output forces are known, its mechanical advantage can be calculated.（表条件）
对于任何机器来说，如果知其输入力和输出力，就能求出其机械效率。

例 12：I'll try to get an illustrated dictionary dealing with technical glossary, which will enable me to translate scientific literature more exactly.（表目的）
我要设法弄一本有插图的技术名词词典，以便把科学文献译得更准确。

Translation at Sentence Level

I. Translate the following sentences into Chinese.

1. An electronic virus is a program that copies itself by taking control of a computer's internal structure.

2. At present, there still exist complex computations in science and engineering which people are unable to make.

3. The reader who is impatient to get on to digital systems should realize that many digital systems also require analog technology to function.

4. Such situations require the use of a carrier signal whose frequency is such that the voice signal will travel through the channel.

5. Microwaves, which have a higher frequency than ordinary radio waves, are used routinely in sending thousands of telephone calls and television programs across long distances.

II. Translate the following sentences into English.

1. 云计算是一个头部公司间竞争相当激烈且技术进步非常快的领域。

2. 有人预测，将来会生产出小到可以植入人脑的计算机。

3. 我们离机器人包揽大量家务活的日子不远了。

____

____

4. 控制器是计算机的一个重要部分，它能使计算机按照人们的意愿进行工作。

____

____

5. 计算机在工程技术领域已经获得了广泛应用，它能使人们从复杂的测量和计算劳动中解脱出来。

____

____

Translation at Paragraph Level

English to Chinese Translation

①Huawei Cloud fully understood Dragonest requirements and provided a tailored cloud migration solution, by selecting specific cloud services to match those needs. ②Onsite technical support was provided during the entire process to ensure Dragonest was familiar with purchased resources and services from Huawei Cloud. ③This allowed Dragonest to focus on its business and save time. ④Huawei Cloud provided Dragonest with a solution with strong performance and scalability to data storage issues, and provided excellent services to reduce solution application obstacles, which enhanced service efficiency. ⑤Huawei Cloud helped Dragonest reduce costs with its reasonable billing mode, and protected its business interest by securing the customer's data security.

本段一共五句话。第一句是一个简单句，主语是 Huawei Cloud，谓语部分由两个并列的动词 understood 和 provided 所带的短语组成，介词 by 后接动名词短语构成一个方式状语，在翻译时需要根据汉语句子的表达习惯调整顺序，置于 provided 所带的动词短语的前面。因此本句可以译为："华为云完全理解 Dragonest 的需求，并通过选择特定的云服务——提供量身定制的云迁移解决方案来满足这些需求。"

第二句是一个复合句，主句的主谓部分是由被动语态构成，在翻译时注意考虑汉语表达中多主动句、少被动句的特点，需要增加谓语动词的发出者"华为云"作汉语句子的主语。此外，作目的状语的动词不定式 to ensure 后跟了一个主系表结构的宾语从句。因此本句可以译为："华为云全程提供现场技术支持，以确保 Dragonest 熟悉购自华为云的资源和服务。"

第三句是一个主谓宾结构较短的简单句，可以顺译为："这使得 Dragonest 能够专注于自身的业务并节省时间。”第四句是一个复合句，主句部分的主语仍然是段落核心主题词 Huawei Cloud，谓语部分是由两个并列的动词 provided 后跟名词短语组成，which 引导一个非限定性定语从句表示主句的结果，全句可以译为：“华为云为 Dragonest 提供了一个性能强大、可扩展性强的解决方案来解决数据存储问题，并提供了优质服务来减少方案应用中的障碍，这些措施提高了服务效率。”第五句与第一句的结构较为相似，可以使用同样的方法翻译为：“华为云通过其合理的计费模式帮助 Dragonest 降低了成本，并通过确保客户数据安全保护了 Drgonest 的商业利益。”

III. Translate the following paragraph into Chinese.

Cloud-based services and businesses that have captured large amounts of personal data emerged as new targets for cyberattacks in China last year, according to a report released by the country's top Internet security risk monitoring authority on Tuesday. IP addresses of major cloud computing platforms in China account for 7.7% of total websites, but the sector was targeted by 53.7% of all online malware threats in 2018, said the report released by the National Computer Network Emergency Response Center. Also, more than half of websites that were threatened by denial-of-service attacks, backdoor programs, or tampering with data and content, provide cloud-based services, the report said.

____

____

____

____

____

____

Chinese to English Translation

①金山云创立于 2012 年，是中国排名前三的互联网云服务商，2020 年 5 月在美国纳斯达克上市，业务范围遍及全球多个国家和地区。②自成立以来，金山云始终坚持以客户为中心的服务理念，提供安全、可靠、稳定、高品质的云计算服务。③金山云依托金山集团 30 年企业级服务经验，坚持技术立业，逐步构建了完备的云计算基础架构和运营体系，并通过与人工智能、大数据、物联网、区块链、边缘计算、AR/VR 等先进技术有机结合，提供超过 120 种适用于政务、金融、医疗、教育、传媒、工业、视频、游戏、电商零售、地产、能源、农业等行业的解决方案，服务 243 家头部客户。④当前，金山云已经在北京、上海、广州、杭州、扬州、天津等国内地区，以及美国、俄罗斯、新加坡等国际区域设有绿色节能数据中心及运营机构。⑤未来，金山云将持续立足本土、放眼国际，通过构建全球云计算网络，联通更多设备和人群，让云计算的价值惠及全球。

本段一共五句话。第一句主要介绍了金山云的发展历史和现有地位，前三个分句有一个共同的主语“金山云”，可以将其确定为英文句的主语，在三个动词“创立”“是”“上市”中确定“是”作为英文句的谓语动词，其他两个转换为非谓语动词，并可以改变原来的语序调整到一起；最后一个分句的逻辑主语“业务范围”与全句的主语不一致，则可以转换为一个独立主格结构。因此第一句可以译为：“Kingsoft Cloud, founded in 2012 and listed on NASDAQ in May 2020, is one of the top three Internet cloud service providers in China, with its business scope covering many countries and regions around the world.”。

第二句可以从主干结构入手，先确定“金山云”作主语，“坚持”“提供”作并列的谓语动词，后接的名词短语可以作宾语，再考虑时态方面，“自……以来”这个时间状语对应的谓语动词应该用现在完成时，因此第二句可以译为：“Since its establishment, Kingsoft Cloud has always adhered to the customer-oriented service concept and provided safe, reliable, stable, and high-quality cloud computing services.”。

第三句介绍金山云的业务，内容较多，句子较长，从句子结构角度分析可以把句子主干结构简化为：“金山云依托……，坚持……，构建……，通过与……结合，提供……。”因“金山云”是全段的核心主题词，翻译时可以把其作为英文句的主语，把“构建”作为谓语动词，并把“依托”“坚持”这两个有动作含义的词转化为名词，最后两个分句则译为一个非限定性定语从句，这样全句可以译为：“Under the support of Kingsoft Corporation with 30-year experience in enterprise services and its own persistence in technological innovation, Kingsoft Cloud has gradually established a comprehensive suite of cloud computing infrastructure and operating system, which makes it possible to provide more than 120 kinds of solutions for many industries, including government administration, finance, healthcare, education, media, industrial engineering, video, game, e-commerce, real estate, energy, agriculture, etc. by integrating with such advanced technologies as AI, big data, IoT, blockchain, edge computing, and AR/VR.”。

第四句介绍金山云的国内外布局，句子结构比较清晰，可以直接译为：“At present, Kingsoft Cloud has set up green and energy-saving data centers and operation offices in Beijing, Shanghai, Guangzhou, Hangzhou, Yangzhou, Tianjin and other domestic regions, as well as in the United States, Russia, Singapore, and other international regions.”。

第五句介绍金山云的未来发展计划，翻译时仍可以把“金山云”作主语，把“持续”作谓语。需要注意的是，本句的时态要用一般将来时，其他部分可用非谓语动词形式来翻译带有动作意味的动词短语，因此本句可以译为：“In the future, Kingsoft Cloud will continue focusing on the domestic market and going global, and extend the value of cloud computing worldwide by building a global cloud computing network and connecting more devices and people.”。

IV. Translate the following paragraph into English.

云计算是一种按需使用的计算机系统资源使用模式。用户只需有一台电脑和网络，就可以进入计算资源共享池（资源包括网络、服务器、存储、应用软件等），发送需求指令给提供云计算的服务商，服务商通过大量的服务器进行计算后，将结果返回给用户。我们常用的搜索引擎就是云计算的一个例证。近两年来，大家使用的云存储也是云计算的服务类型之一，用户无须拥有容量超大的计算机硬盘，只需申请一个云存储平台的账号，就可以把文件存储在服务商提供的云端存储平台上。据报道分析，从云平台上发出的攻击增多，是因为云服务存在便捷性、可靠性、低成本、高带宽和高性能等特性，并且云网络流量的复杂性有利于攻击者隐藏真实身份，更多地利用云平台设备作为跳板或控制端发起网络攻击。

____

____

____

____

____

____

Workshop

I. Choose one cloud computing service you enjoy most in daily life and do research on how it is developed. Then write an outline of your research results and make a five-minute oral presentation to the class.

Service: ______________________________

Company: ______________________________

Functions: ______________________________

Main Features: ______________________________

Reasons for Recommendation: ______________________________

II. Fill in the table below to make a comparison between two of the world's top 10 companies in cloud computing and discuss with your partners about which one is better.

Details	Company A	Company B
Name		
Country/Region		
Headquarters		
Number of Employees		
Market Share		
Annual Revenue		
Business Model		
Advantages		
Disadvantages		
Prospects		

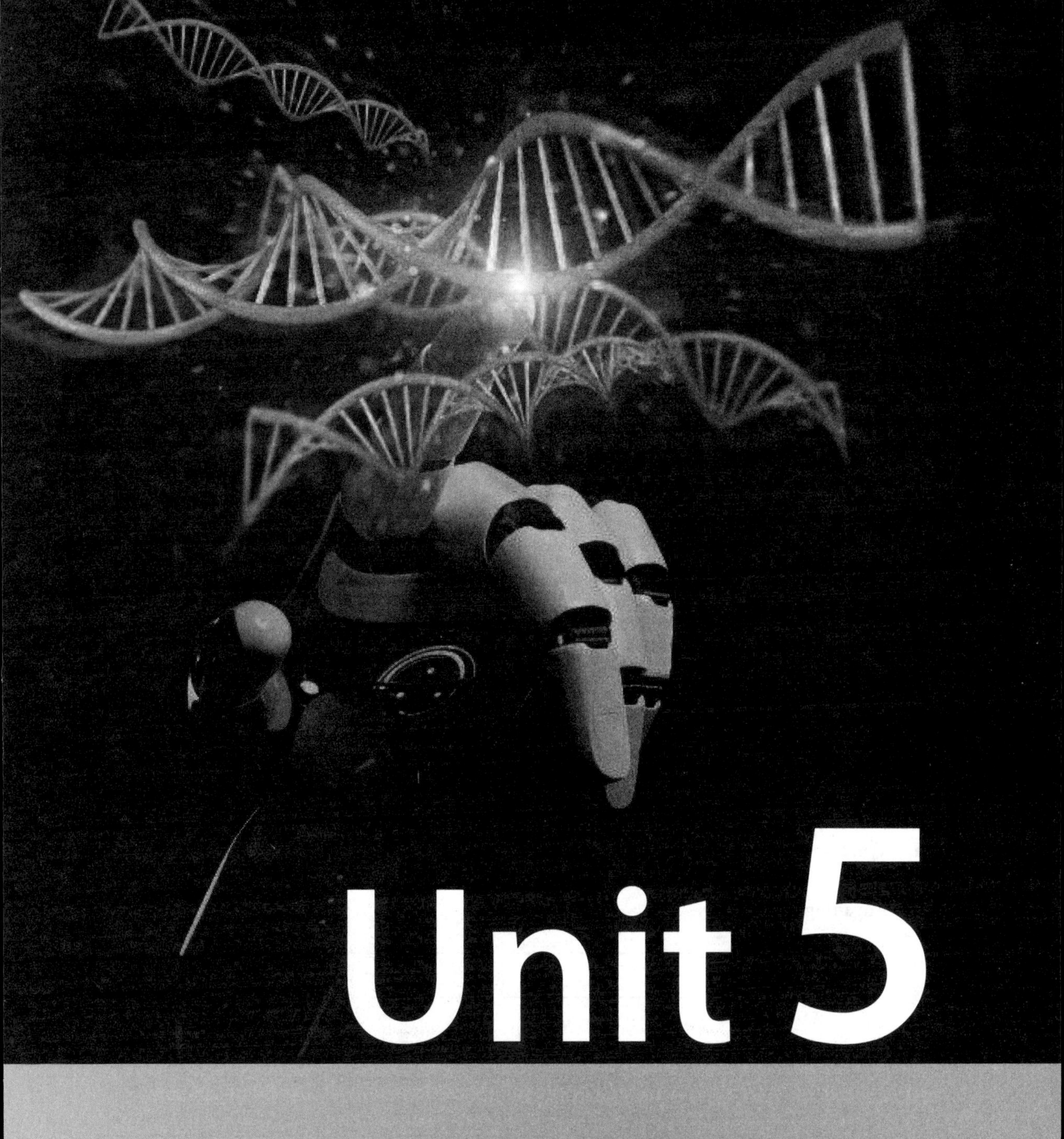

Unit 5

Artificial Intelligence and Internet of Things

Learning Objectives

In this unit, you will learn:

- the combination of Internet of Things and artificial intelligence;
- the applications of Internet of Things in real life;
- major IoT risks that accompany business today;
- writing skills—the usage of parallelism;
- translating skills—amplification and omission.

Lead-in

I. Discuss the following questions with your partners.

1. Do you know anything about Internet of Things?
2. How does Internet of Things impact people's daily life? Can you give an example?
3. What is the definition of Internet of Things? Why are there so many definitions of it?
4. What are the characteristics of Internet of Things?

II. Work in pairs to discuss the advantages and disadvantages of Internet of Things.

Advantages:

1. ______________________________
2. ______________________________
3. ______________________________

Disadvantages:

1. ______________________________
2. ______________________________
3. ______________________________

Realizing the Future and Full Potential of Connected IoT Devices with AI

1 Business leaders today need to be reinventors with a clear vision to integrate IoT with AI across their business. All around us in the world, the **convergence** of new technologies and intelligent machines is turning things smarter and reshaping the future of enterprises. **Leading-edge** companies, both new and old, are leveraging emerging technologies to automate and **optimize** their business performance.

2 The excellent business leaders are capturing all forms of data from various devices that are joined together. They are using AI-based applications to plumb that data to reach new levels of operational and financial proficiency. They are reinventing industries, extracting valuable insights from the **proliferating** data, and creating new opportunities and markets for their business. To innovate their products and services, the reinventors distinguish themselves from others by using customer data and analyzing competitors. They **scrutinize** any available data to understand customer needs and anticipate customer demands. They also analyze their competitors' responses to speculate what customers want.

3 The reinventors know that the development of IoT data is twice faster than that of social and computer-generated data. As a result, we can see that they are moving from simply capturing IoT sensor data to leveraging it for competitive advantages.

The Foundations of IoT and AI

4 Beyond doubt, the early applications of IoT are delivering great value for many companies. For example, they reshape customer experience by putting consumers into context and offering new **avenues** for **engagement**. At the same time, the reinventors understand that along with AI, the full potential of IoT can be fully used for business model reinvention. With AI, they are introducing new products, discovering new opportunities, reducing risks, and increasing **revenue**. When designed, traditional methods of analyzing structured data were not set to efficiently deal with the vast amounts of real-time data collected from IoT devices. It is critical of AI-based analysis and response to extract optimal value from that data. To lead in today's business, reinventors need to

have a vision to integrate IoT with AI across their business to foster deeper customer relationships, find new sources of value through data, and accelerate the digital transformation of their business operations.

AI and Machine Learning

5 In essence, IoT is a data challenge. The traditional approach to programmable computing is **shepherding** data through a series of pre-determined, if-then-else processes to arrive at outcomes. Such a way is unable to process the degree and kind of data needed to fulfil the true promise of IoT. But with AI and machine learning, systems don't need to be explicitly programmed. They learn from interactions with users and experiences with the environment, which enables them to **keep pace with** IoT complexity and **havoc**. Meanwhile, machine learning is moving from computer labs and Web applications to the physical world, where current levels of digital training data and computing power make it functional and actionable. This ability to **embed** learning capabilities within the IoT device itself, plus marrying device-centric insights with **aggregated** intelligence in the cloud, is expected to dramatically improve outcomes.

Putting AI and IoT into Action

6 So, how does the combination of IoT and AI play out in the industrial situations?

7 When a clothing retailer is to improve the customer experience, traditionally he may work to gather data about online shopping habits. But he may find it is far more difficult to quantify in-store behavior. With IoT and AI, a store can combine traditional sources of structured data (such as supply chain, **inventory**, RFID tags, and point of sale) with new sources of less-**quantifiable** information (such as in-store foot traffic, social media, and even weather data) to get a more complete understanding of customer behavior. As a result, the clothing retailer can **correlate** the data, identify patterns, and make specific, **unbiased** recommendations on everything from store layout and **merchandising**, to supply chain management and product design.

The Journey Towards Intelligent IoT and Automation

8 Combining IoT with AI can be the aspiration to many industries, even those with a mature IoT network. To fully **capitalize on** this ambition, there are several necessary steps as recommended in the following:

- Plan a connected, software-driven world. Develop and **articulate** a clear vision of AI/IoT strategy, **back** it **up** with reinvention roadmaps and execution plans, and then communicate this vision with the company's **stakeholders**.
- Create an organizational culture of inclusion. Promote and support collaboration and knowledge-sharing across employees.

- Equip engineers, developers, and operations to deal with unprecedented complexity and technology developments. Prepare **agile** and distributed teams to help deliver the right skills.
- Take advantage of massive internal and external data across the ecosystem, and use the data to design new customer experience. Evaluate and include select competitors to innovate products and services.
- **Infuse** operations and customer experience with IoT intelligence and automation as AI can enable new classes of IoT products and services that sense, reason, and learn.
- Optimize processes to improve quality and operational efficiency with less human and financial risk.
- Exploit new revenue sources by transforming products, services, and experiences with individualized interactions delivered through new services, great design, and new features that customers cherish.

9 And remember, it is a journey.

Notes

if-then-else a high-level programming language statement that compares two or more sets of data and tests the results. If the results are true, the THEN instructions are taken; if not, the ELSE instructions are taken.

RFID the abbreviation for radio-frequency identity (or identification).

Words and Expressions

convergence /kən'vɜːdʒəns/ *n.* the process by which different ideas, groups, or societies become similar or the same 趋同性

leading-edge /ˌliːdɪŋ 'edʒ/ *adj.* the most important and advanced (machine, system, etc.) （机器、系统等）最现代的，最先进的

optimize /'ɒptɪmaɪz/ *v.* to improve the way that something is done or used so that it is as effective as possible 使最优化，使尽可能完善；使尽量有效

proliferating /prə'lɪfəreɪtɪŋ/ *adj.* increasing quickly in number or amount 激增的

scrutinize /'skruːtənaɪz/ *v.* to examine somebody/something very carefully 仔细查看，认真检查，细致审查

avenue /'ævənjuː/ *n.* a possible way of achieving something 方法，途径

engagement	/ɪnˈgeɪdʒmənt/	*n.*	being involved with somebody/something in an attempt to understand him/her/it 参与
revenue	/ˈrevənjuː/	*n.*	the money that a government receives from taxes or that an organization, etc. receives from its business 财政收入；税收收入；收益
shepherd	/ˈʃepəd/	*v.*	to lead or guide a group of people somewhere, making sure that they go where you want them to go 带领（一群人）；引导（人群）
keep pace with			to change quickly in response to something 与……保持同步
havoc	/ˈhævək/	*n.*	a situation in which there is a lot of damage or a lack of order, especially when it is difficult for something to continue in the normal way 混乱；灾难
embed	/ɪmˈbed/	*v.*	to put something firmly and deeply into something else, or to be put into something in this way （使）嵌入，把……插入
aggregate	/ˈægrɪgeɪt/	*v.*	to put together different items, amounts, etc. into a single group or total 总计，合计
inventory	/ˈɪnvəntri/	*n.*	all the goods in a shop 存货，库存
quantifiable	/ˌkwɒntɪˈfaɪəbl/	*adj.*	able to be measured or counted in a scientific way 可计量的
correlate	/ˈkɒrəleɪt/	*v.*	to show there is a close connection between two or more facts, figures, etc. 显示……的紧密联系
unbiased	/ʌnˈbaɪəst/	*adj.*	(information, opinions, advice, etc. that is) fair and not influenced by other people's opinions 不偏不倚的，公正的，无偏见的
merchandising	/ˈmɜːtʃəndaɪzɪŋ/	*n.*	the way in which shops and businesses try to sell their products 推销，展销
capitalize on			to use a situation or something good that you have, in order to get an advantage for yourself 充分利用
articulate	/ɑːˈtɪkjʊleɪt/	*v.*	to express or explain your ideas or feelings clearly in words 清楚地表达
back…up			to support or confirm (an idea or intention) by action （以行动）支持
stakeholder	/ˈsteɪkhəʊldə(r)/	*n.*	a person or company who has invested money in a particular organization, project, etc., or who has some important connection with it, and

			therefore is affected by its success or failure 参与人；投资者；有权益关系者
agile	/'ædʒaɪl/	*adj.*	able to think very quickly and intelligently （思维）机敏的
infuse	/ɪn'fjuːz/	*v.*	to make somebody/something have a particular feeling or quality 使具有；注入（某特性）

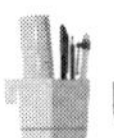

Useful Terms

supply chain	供应链
point of sale	销售点
in-store foot traffic	店内人流量

Exercises

Comprehension Check

I. Answer the following questions according to the text.

1. What benefits can leading-edge enterprises obtain if they integrate IoT with AI?
2. Why does IoT with traditional methods of data analysis fail to reach the full potential?
3. What is the traditional approach to programmable computing?
4. What results can a clothing retailer get if his store combines IoT with AI?
5. What if engineers and developers run into unprecedented complexity when conducting an AI/IoT strategy?

II. Read the following statements carefully and decide whether they are true (T) or false (F) without turning back to check the text.

1. ______________ The top enterprises are in the leading position because they use IoT data to analyze both customer needs and competitors' responses.
2. ______________ Traditional methods of data analysis can efficiently process the large amount of real-time data from IoT devices.
3. ______________ Machine learning is stepping out of computer labs and entering the

physical world.

4. ______________ Even with the assist of AI, it is difficult for clothing retailers to quantify customers' in-store behavior.

5. ______________ The journey to fully capitalize on the ambition of IoT and automation is still on trial, and the whole staff of a company might get involved.

III. Choose the best answer to each of the following questions according to the text.

1. With the addition of AI, what is the full potential of IoT business model?

   A. Reinventors can introduce new products.

   B. Reinventors can discover new opportunities.

   C. Reinventors can reduce risk and increase revenue.

   D. All of the above.

2. Machine learning stays in tune with IoT complexity and havoc because it ______________.

   A. challenges IoT data

   B. can shepherd through pre-determined processes to arrive at outcomes

   C. can program systems explicitly

   D. can learn from interactions with users and experiences with the environment

3. How can a clothing retailer enhance the in-store customer experience?

   A. By gathering data about customers' online shopping habits.

   B. By quantifying the information on in-store customer behavior.

   C. By correlating traditional structured data with new less-quantifiable information.

   D. It is difficult to enhance in-store service since in-store customer behavior is hard to quantify with data.

4. Which of the following steps is NOT recommended to a company that works to combine IoT with AI?

   A. Focusing on building a mature IoT network.

   B. Backing up AI/IoT idea with execution plans.

   C. Communicating AI/IoT vision with stakeholders and employees.

   D. Evaluating internal and external data to innovate products and services.

5. For today's business leaders, they should extend their vision to ______________.

   A. capture all forms of data from interconnected devices

B. automate their business performance

C. get IoT data closely linked to AI applications

D. reinvent new technologies and intelligent machines

Vocabulary Building

IV. Fill in the following blanks with the words and phrases given in the box. Change the form if necessary.

optimize	proliferate	engagement	shepherd	havoc
keep pace with	correlate	capitalize on	articulate	quantifiable

1. This is yet another example of the growing ______________ of world leaders in health issues.
2. Some past offenders have been released only to cause ______________ and end up in jail again.
3. Attempts to ______________ specific language functions with particular parts of the brain have not advanced very far.
4. How do we best ______________ the role of corporations in alleviating world poverty?
5. What verifiable and ______________ evidence can we show that this is logical?
6. Many people are unable to ______________ the unhappiness they feel.
7. The companies had hoped to take long-term stakes in several of China's biggest banks as a way to ______________ this country's growing economy.
8. Books and articles on artificial intelligence have ______________ over the past several years.
9. All or nearly all visitors will be ______________ by guides to this field and then onto the bus.
10. The speed of technological development is so fast that we could hardly ______________ modern trends.

V. Match the words in the left column with the explanations in the right column.

1. reshape — A. to put two things or groups together so that they can work or act together
2. proficiency — B. to change the structure or organization of something
3. optimal — C. to introduce changes and new ideas in the way something is

done or made

4. combine — D. a good standard of ability and skill

5. innovate — E. the best or most suitable

Word Formation

VI. Fill in the following blanks with the words in capitals. Change the form if necessary. An example has been given.

e.g. *Leading-edge companies, both new and old, are leveraging emerging technologies to automate and optimize their business performance.* **PERFORM**

1. A connected car is able to optimize its own operation and ______________ using onboard sensors and Internet connectivity. **MAINTAIN**

2. Heavy ______________ on one client is risky when you are building up a business. **RELY**

3. We are often attracted to somebody first by his physical ______________. **APPEAR**

4. The building work is creating constant noise, dust, and ______________. **DISTURB**

5. The company provides advice and ______________ in finding work. **ASSIST**

Translation

VII. Translate the following sentences into English with the words and phrases in brackets.

1. 领头羊企业需要利用新兴技术来优化业务绩效。(leverage; optimize)

____

____

2. 人工智能和机器学习使物联网系统能够从与用户的交互中学习，并同步了解复杂、混乱的情况。(interaction; keep pace with)

____

____

3. 为了将人工智能与物联网相结合，企业领导者应该制订并阐明清晰的人工智能/物联网战略愿景。(combine; articulate)

____

____

Applications of Internet of Things in Real Life

1 Smart devices, also called "connected devices", are designed to capture and utilize every bit of data people share or use in everyday life. The data will be used by smart devices to interact with people on daily life and tasks. This new wave of **connectivity** is not just going with laptops and smart phones, but it is going towards connected cars, smart homes, connected wearables, smart cities, and connected healthcare. According to Gartner report, by 2020 connected devices across all technologies will reach 26 billion. To get a **glimpse** of how the applications of IoT will transform people's lives, nine areas are listed down where IoT is much awaited and companies are preparing to surprise people with smart devices.

Smart Home

2 Smart home has become the revolutionary ladder of success in the **residential** spaces. Imagine you could switch on air conditioning before reaching home, or switch off lights after you have left home, or unlock the doors to friends for temporary access even when you are not at home. Smart home products are promised to save not only time and energy but also money, which is the biggest expense in a homeowner's life. Smart home companies like Nest, Ecobee, Ring, and August, to name a few, are becoming household brands and planning to deliver a never-seen-before experience.

Wearables

3 Wearable devices, installed with sensors and software, can collect data and information about users. This data is later pre-processed to extract essential insights about the users. Such wearable devices broadly cover fitness, health, and entertainment requirements. That is why the **pre-requisite** for wearable applications is to be highly energy-efficient, ultra-low power, and small sized. Companies like Google and Samsung have invested heavily in building such devices.

Connected Cars

4 A connected car is a vehicle which, with onboard sensors and Internet connectivity, is able to optimize its own operation, maintenance as well as the comfort of passengers. Most large automakers and some brave **start-ups** are working on connected car solutions such as enhancing

vehicle's internal functions and driver's in-car experience. Major brands like Tesla, BMW, Apple, and Google are exerting efforts to bring the next revolution in automobiles.

Industrial Internet

5 Industrial Internet, also termed as Industrial Internet of Things (IIoT), is **empowering** industrial engineering with sensors, software, and big data **analytics** to create brilliant machines. IIoT holds great potential for quality control and **sustainability**. The supply chain will be efficiently increased by the applications for tracking goods, real-time information exchange about inventory among suppliers and retailers, and automated delivery. The improvement of industry productivity will generate $10 trillion to $15 trillion in GDP worldwide over next 15 years.

Smart Cities

6 With IoT, major problems in big cities like pollution, traffic **congestion**, and shortage of energy supplies will be solved. For example, Smart Belly enabled by **cellular** communication will send alerts to **municipal** services when a bin needs to be emptied. Installed with sensors, free available parking slots across the city can be easily found by citizens with clicks on the Web applications on their phones. Other issues can also be detected by the sensors, such as meter **tampering**, general **malfunctions**, and any installation issues in the electricity system.

Smart Farming

7 Smart farming is one of the fields growing fast in IoT. Governments are encouraging and helping farmers to increase food production with advanced techniques. From the data, farmers are getting meaningful insights to yield better return on investment. Some uses of IoT are sensing for soil moisture and **nutrients**, controlling water usage for plant growth, and determining custom fertilizer. Another use of IoT in agriculture is livestock monitoring, which brings animal **husbandry** and cost saving. Gathering data about the health and well-being of the chicken using IoT applications, ranchers can know early about the sick animals, pull them out, and prevent a large number of sick ones, which consequently will increase the poultry production.

Smart Retail

8 IoT provides an opportunity for retailers to connect with the customers and enhance their in-store experience. Smart phones will be the way for retailers to remain connected with their customers even out of store. Retailers can serve their consumers better by interacting through smart phones and using Beacon technology. They can also track consumers' path through a store to improve store layout and place **premium** products in high-traffic areas.

Smart Grids

9 Power grids of the future will not only be smart enough but also be highly reliable. The basic idea behind the smart grids is to collect data in an automated fashion and analyze the behavior

of electricity consumers and suppliers for improving efficiency as well as economics of electricity use. Smart grids will also be able to detect sources of power outages more quickly at individual household levels like nearby solar panel, making a distributed energy system possible.

Connected Healthcare

10 Connected healthcare yet remains the sleeping giant of the IoT applications. IoT in the healthcare industry is aimed at empowering people to live a healthier life by wearing connected devices. The collected data will help in personalized analysis of an individual's health and provide **tailor-made** strategies to combat illness.

11 The future of IoT will be more fascinating where billions of things will be talking to each other and human intervention will become least. IoT will bring a macro shift in the way we live and work.

Notes

Nest also called Nest Labs, an American manufacturer of smart home products including thermostats and smoke detectors, and security systems including smart doorbells and smart locks.

Ecobee a Canadian home automation company that makes thermostats for residential and commercial use.

Ring formerly Doorbot, a global home security company that manufactures a range of home security products that incorporate outdoor motion-based cameras and doorbells.

August a home automation company in San Francisco, focusing on Wi-Fi connected door locks and doorbell cameras.

Tesla an American automotive and energy company that specializes in electric car and solar panel manufacturing.

BMW a German multinational company which currently produces automobiles, motorcycles, and aircraft engines.

Apple an American multinational technology company that designs, develops, and sells consumer electronics, computer software, and online services.

Smart Belly originally a solar-powered, rubbish-compacting bin, manufactured by the U.S. company Bigbelly. It is put in use in public spaces such as parks, beaches, amusement parks, universities, retail properties, grocery industry, and food service operators.

Beacon technology used by marketers to better personalize the messaging and mobile ads based on the customers' proximity to their retail outlet.

Words and Expressions

connectivity	/kəˌnekˈtɪvəti/	*n.*	the state of being connected or the degree to which two things are connected 连接（度）；联结（度）
glimpse	/ɡlɪmps/	*n.*	a short experience of something that helps you to understand it 短暂的感受，体验，领会
residential	/ˌrezɪˈdenʃl/	*adj.*	relating to homes rather than offices or businesses 家庭的；住宅的
prerequisite	/ˌpriːˈrekwəzɪt/	*n.*	something that must happen or exist before something else is possible 先决条件
start-up	/ˈstɑːt ʌp/	*n.*	a small business that has recently been started by someone 新企业
empower	/ɪmˈpaʊə(r)/	*v.*	to give somebody the power or authority to achieve something 使能够
analytics	/ˌænəˈlɪtɪks/	*n.*	a careful and complete analysis of data using a model, usually performed by a computer; information resulting from this analysis 分析学；解析学
sustainability	/səˌsteɪnəˈbɪləti/	*n.*	the property of being sustainable 持续性，永续性，能维持性
congestion	/kənˈdʒestʃn/	*n.*	a place which is extremely crowded and blocked with traffic or people 拥塞
cellular	/ˈseljələ(r)/	*n.*	connected with a telephone system that works by radio instead of wires （电话系统）蜂窝状的，蜂窝式的
municipal	/mjuːˈnɪsɪpl/	*adj.*	relating to or belonging to the government of a town, city, or district 市政的，市办的
tamper	/ˈtæmpə(r)/	*v.*	to make changes to something without permission, especially in order to damage it 胡乱摆弄；擅自改动（尤指有意损坏）
malfunction	/ˌmælˈfʌŋkʃn/	*n.*	a fault in the way a machine or part of someone's body works （机器的）故障，失灵；（人体器官的）功能障碍
nutrient	/ˈnjuːtriənt/	*n.*	a chemical or food that provides what is needed for plants or animals to live and grow 养分，营养物
husbandry	/ˈhʌzbəndri/	*n.*	farming, especially when done carefully and well 农牧业；饲养业
premium	/ˈpriːmiəm/	*adj.*	of very high quality 优质的
tailor-made	/ˌteɪlə ˈmeɪd/	*adj.*	specially designed for a particular person or purpose 专门设计的

Useful Terms

ultra-low power　　超低功耗

custom fertilizer　　定制肥料

high-traffic area　　人流量密集的区域

power grid　　电网

power outage　　停电，断电

Exercises

Comprehension Check

I. Identify the paragraph from which the information is derived.

1. ______________ People can open the door to a friend for temporary visit when they are not at home.
2. ______________ Connected devices can help control quality, track goods, and perform automated delivery.
3. ______________ When a dustbin is full, a smart device will alert municipal services to empty the bin.
4. ______________ Consumers' foot path through a store can be tracked to help improve store layout and product placement.
5. ______________ Companies can get data and information about people's fitness, health, and entertainment requirements.
6. ______________ Connected devices may offer data and information about soil moisture to farmers and help control water usage.
7. ______________ The users' behavior and consumption of electricity can be analyzed and electricity efficiency can be improved.
8. ______________ People's operation and comfort while driving cars may be optimized with onboard sensors and Internet connectivity.
9. ______________ Ranchers may separate the sick cattle and prevent a large number of sick ones in advance.

10. ______________ The new wave of connectivity may involve devices up to 26 billion by 2020.

II. Answer the following questions according to the text.

1. Do you agree that IoT has made a big impact on people's daily life? What impact has it made?
2. Among the applications listed in the text, which do you think customers are more likely to use? Can you rank these applications according to the statistics?

Vocabulary Building

III. Fill in the following blanks with the words and phrases given in the box. Change the form if necessary.

start-up	switch on	empower	sustainability	congestion
municipal	get a glimpse of	nutrient	premium	residential

1. In an attempt to reduce ______________, the U.S. has had car pool lanes for decades.
2. The model of the American city dominates ______________ complexes and shopping centers with its office blocks.
3. A pre-set code is given to the farmer to ______________ or off the pump set for feeding animals.
4. For now, the only bright spots in the labor market are small businesses and high-tech ______________.
5. ______________ tobacco products from all over the world build the core of the brand Davidoff.
6. You must delegate effectively and ______________ people to carry out their roles with your full support.
7. What people love most about smart home is that they can ______________ a different way of life.
8. Their votes were tallied by hand at the precinct, ______________, provincial, and finally national levels.
9. Reducing wastage, for example, by buying products with little packaging, is a key component of ______________.
10. The U.S. law does not require labels unless the composition or ______________ value of the product is significantly changed.

Text C

Major IoT Risks in Business Today

1 Do you remember the 2016 **DDoS attack**? Considered as one of the most complex and **devastating** cyberattacks, it took down major websites like Twitter and greatly worsened user experience on numerous famous sites including PayPal, Spotify, CNN, Mashable, and Yelp. The attack was particularly astonishing due to the enormous **traffic** coming from millions of unique Internet addresses. The mess was made by a hacker who had identified a **loophole** in a certain model of security cameras. By taking over these cameras, the attacker directed massive traffic to targeted websites, which caused them inaccessible to **legitimate** users. Though the Internet of Things has multiple advantages, this attack is a typical example of how IoT turns to be a weapon of mass destruction in the hands of **malicious** people. And it's not just cameras that can be hijacked in this way, but anything with an Internet connection could be a potential target for hackers, from cars and refrigerators to thermostats and smart locks.

> 分布式拒绝服务攻击
> *adj.* 毁灭性的
> *n.* 信息流量，通信量
> *n.* 漏洞
> *adj.* 合法的
> *adj.* 恶意的

2 Today the Internet is no longer a network of traditional computing devices like servers, routers, desktops, laptops, tablets, and smart phones. IoT has actually become a convenience in many ways as the number of IoT gadgets is expected to eventually far exceed **conventional** computing gadgets. For example, refrigerators can relay an update of the freshness of food; a car can transmit oil level information to its owner. Yet, as is shown in several IoT-related cyberattacks, **formidable** risks that come with the convenient gadgets cannot be ignored. Some of the biggest IoT risks are covered below.

> *adj.* 传统的；常见的
> *adj.* 可怕的

Risks in the Manufacturing Process of IoT Devices

3 Manufacturer **omission** is responsible for the vast majority of security issues **bedeviling** IoT devices. Manufacturers release an

> *n.* 疏忽；遗漏
> *v.* 困扰

untold number of IoT devices into the market each day, many of which are new models and have undiscovered vulnerabilities. Since many device manufacturers see Internet connectivity as a plus to their device functions rather than a core feature, they do not devote as much time and resources as they should to ensuring that their products are secure from cyberattacks.

adj. 数不清的

4 The biggest IoT risks **emanating** from the manufacturing process include weak passwords, unsecured hardware, absence of a **patching mechanism**, and insecure data storage. For instance, some smart refrigerators expose Gmail **credentials**. Some fitness trackers with Bluetooth connectivity remain visible after their first-ever pairing. Though there isn't a universal standard for securing IoT devices, that is not a justifiable reason for creating poorly secured devices.

v. 产生
修补机制
n. 认证信息

Lack of User Awareness and Knowledge

5 While the majority of the biggest IoT risks can be traced to the manufacturing process, users are far more dangerous drivers of IoT security risks. Even though the average Internet users, thanks to decades of awareness, are fairly adept at avoiding **phishing** e-mails, disregarding suspicious attachments, running virus scans on their computer, and creating a strong password, IoT remains a new territory to common users, even for many **seasoned** IT professionals. This is especially so when users are ignorant of IoT **functionality**. In fact, a human is often the easiest to deceive when hackers **infiltrate** a restricted network without raising any suspicion.

n. 网络钓鱼，网络欺诈
adj. 经验丰富的
n. 功能，功能性
v. 潜入

6 Take the 2010 Stuxnet worm attack as an example. The attack on an Iranian nuclear facility was caused by the infection of **centrifuge**-controlling software via a USB flash drive plugged into one of the plant's computers. Modern centrifuges are a type of IoT device as they are heavily IT-dependent. It was estimated that Stuxnet physically damaged about 1,000 centrifuges.

n. 离心机

Difficulty in Patching and Updating Management

7 Patches and updates are necessarily needed to keep IoT devices secure and should be applied as soon as they are released. But the nature and use of IoT devices don't always make them easy to update regularly—**if at all**. Think about the sensors that spread across

如果某事真会发生的话

hundreds of acres on a farmland, or the IoT devices on a factory floor that cannot be taken offline for updates without hugely impacting production. Worse still, even where patches can be applied regularly, there's often no means for the user to **rollback** changes to the last known good state in the event that an update leads to software corruption or instability. No matter how much work a manufacturer puts into creating secure hardware and software for its IoT product, new vulnerabilities will inevitably be discovered at some point in the future.

v. 还原；退回

The Physical Insecurity of IoT Devices

8 IoT devices commonly run with little or no human intervention. Sometimes, these devices are installed in remote locations where they may stay for weeks or months without anyone physically checking on them. Such isolation leaves them in great danger of theft or physical tampering. Criminals might steal the device or use a flash drive to introduce **malware**. They might also interfere with the functioning of the IoT device rendering the data it collects and relays unreliable.

n. 恶意软件

9 IoT is bringing efficiency to everyday processes. However, it still has numerous security and risk challenges and even more will emerge in the future. As the diversity of IoT devices grows, so will the complexity of the security challenge. To reap the benefits of IoT, keeping these devices secure by **mitigating** against the biggest IoT risks is **paramount**.

v. 减轻
adj. 首要的

Exercises

Comprehension Check

I. Answer the following questions according to the text.

1. How did the famous websites like Twitter and PayPal get attacked in 2016?
2. What causes the vast majority of IoT security problems?
3. Why is it not easy to regularly patch and update the software of IoT devices?

4. What are the physical dangers faced by IoT devices?

5. What opinions does the author mainly express?

II. Read the following statements carefully and decide whether they are true (T) or false (F) without turning back to check the text.

1. ______________ IoT devices are no longer helpful gadgets because they have become hackers' destructive weapons.

2. ______________ Both manufacturers and users should be responsible for the cyberattacks through IoT devices.

3. ______________ Since hackers keep on finding new loopholes in IoT devices, patching and updating turn to be necessary.

4. ______________ When IoT devices are installed in remote locations, they can avoid being attacked or stolen.

5. ______________ As people enjoy the benefits brought by IoT, it is vital to keep the devices away from numerous cyberattacks.

III. Discuss the following questions based on your understanding of Internet of Things.

1. Do you think there are any differences between the risks caused by IoT and AI? If yes, explain it with examples; if not, share your reasons.

2. Can you describe one of the IoT risks in a certain area (i.e. the risks in smart home or in healthcare)?

3. How can people protect IoT devices from security problems? Please elaborate it from the perspective of developers and users respectively.

Writing Skills

排比

作为一种修辞方法，排比（parallelism）是指使用相似的单词、短语或句子来表明具有相同重要性的内容。这种并列结构使书面表达更为自然，也更具有可读性，还增加了书面作品的对称性、有效性和平衡性。演讲中也常使用排比句来说服、激励或感染听众，这有助于引起他们的注意，并帮助他们理解复杂的内容。对于英语母语者来说，使用排比通常是一种习惯。他们会说“I like reading, writing, and painting.”，不会说“I like to read, writing, and painting.”。英语学习者常犯的一个错误是没有使用一致的并列结构，比如“Writers can use an online dictionary to find help with these issues: word meanings, pronunciations, and finding correct spellings.” 中 words meanings、pronunciations 和 finding correct spellings 不是一致的并列结构，应改为“Writers can use an online dictionary to find help with these issues: word meanings, pronunciations, and correct spellings.”。

并列结构一般包括非谓语形式并列、名词和名词短语并列（也称主语并列、宾语并列、表语并列）、谓语动词并列、补语并列、被动句句式并列、从句并列与完整句式并列。

例 1：Mother was very busy gathering the laundry, dusting the furniture, and washing the dishes.

例 2：To succeed in life, you need to take advantage of opportunities and to follow your dreams.

例 3：He likes television shows that have deep characters, interesting stories, and good actors.

例 4：We are giving away our furniture, selling our house, and moving to Spain.

例 5：My fellow citizens: I stand here today humbled by the task before us, grateful for the trust you have bestowed, and mindful of the sacrifices borne by our ancestors.

例 6：My face is washed, my hair is combed, and my teeth are brushed.

例 7：My fellow Americans, ask not what your country can do for you; ask what you can do for your country.

例 8：I have a dream that my four little children will one day live in a nation where they will not be judged by the color of their skin but by the content of their character. I have a dream today.

Underline the parallelism in the following sentences and figure out their structures.

e.g.: *The reinventors are capturing all forms of data from a variety of interconnected devices. They are reinventing industries, extracting valuable insights from the proliferating data, and creating new opportunities and markets for their business.* (并列谓语)

1. The leading-edge companies excel at using customer data to analyze their competitors, scrutinize available data, and anticipate what customers want.

____

____

2. You should equip engineers, developers, and operations to deal with unprecedented complexity and technology developments.

____

____

3. AI can enable new classes of IoT products and services that sense, reason, and learn.

____

____

4. Develop new revenue sources by transforming products, services, and experiences with individualized interactions.

____

____

5. Individualized interactions are delivered through new services, great design, and new features that customers cherish.

____

____

Translating Skills

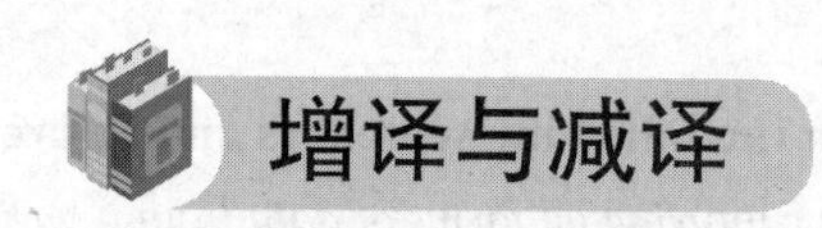

增译与减译

增译（amplification），也叫增词译法，指翻译时在原文基础上添加必要的单词、词组、分句或完整句，使译文在语法和语言形式上符合译文的表达习惯，并在文化背景、词语联想方面与原文一致，从而使译文与原文在内容、形式和精神三方面都对等起来。

例 1：This great scientist was born in New England.
这位伟大的科学家出生于美国东北部的新英格兰。（增加定语，提供背景知识）

例 2：To get a glimpse of how the applications of IoT will transform people's lives, nine areas are listed down where IoT is much awaited and companies are preparing to surprise people with smart devices.

为了了解物联网应用如何改变人们的生活，我们列出了期待物联网的九个领域，在这些领域中，企业正准备用智能设备给人们带来惊喜。（增加主语、介词短语）

例 3：The traditional approach to programmable computing simply cannot process the degree and kind of data needed to fulfill the true promise of IoT. With AI and machine learning, systems don't need to be explicitly programmed.

传统的可编程计算方法根本无法处理实现物联网真正所需的数据程度和类型。现在有了人工智能和机器学习，系统不再需要显式编程。（增加时间状语）

减译（omission），也叫省词译法，指在英译汉过程中可以省略原文中一些词语的做法，一般指那些在译文中保留下来反而使行文显得累赘、啰唆且不符合汉语表达习惯的词语。减译一般用于以下两种情况：

第一，从语法角度进行减省。由于英汉两种语言在语法上的差异，英语中使用较多的代词、冠词、介词和连词在翻译时可以略去不译，从而使译文更加通顺、流畅。此外，英语中的谓语动词有时在汉语中也可省略。

例 4：They are analyzing their competitors' responses to customer demands.

他们正在分析竞争对手对客户需求的反应。（省略代词）

例 5：They also understand that with the addition of AI, the full potential of IoT can be leveraged for business model reinvention.

他们也明白，随着人工智能的加入，物联网的全部潜力可以用于商业模式的再造。（省略连词、冠词）

例 6：When the pressure gets low, the boiling-point becomes low.

气压低，沸点就低。（省略动词）

第二，从修饰角度进行减省。有些词语在英语里是必不可少的，但在汉语中却并非如此，如果直译成汉语通常显得累赘、啰唆。为了使译文简洁、通畅，以下三种情况在翻译时可以进行省略或精简：一是在译文中可有可无或多余的词语；二是意思已包含在上下文的词语；三是含义在译文中不言而喻的词语。

例 7：They scrutinize any available data to understand customer needs, and anticipate what customers want.

他们仔细检查所有可用的数据，以了解并预测客户的需求。（意思已包含）

例 8：They also know that Internet of Things data is growing twice as fast as social and computer-generated data. As a result, we can see that they are moving from simply capturing IoT sensor data to leveraging it for competitive advantages.

他们还知道物联网数据的增长速度是社交和计算机生成数据的两倍。因此他们正在从简单地捕获物联网传感器数据转向利用它来获得竞争优势。（简洁明确）

Translation at Sentence Level

I. Translate the following sentences into Chinese.

1. Smart home has become the revolutionary ladder of success in the residential spaces, and it is predicted to be as common as smart phones.

____

____

2. Smart grids will also be able to detect sources of power outages more quickly at individual household levels like nearby solar panel, making a distributed energy system possible.

____

____

3. Do you remember the 2016 DDoS attack, which was considered as one of the most complex and devastating cyberattacks?

____

____

4. Though the Internet of Things has multiple advantages, this attack is a typical example of how IoT turns to be a weapon of mass destruction in the hands of malicious people.

____

____

5. Patches and updates are necessarily needed to keep IoT devices secure and should be applied as soon as they are released.

____

____

II. Translate the following sentences into English.

1. 自从电脑普及以来，大多数人都意识到需要确保它的安全性。

____

____

2. 我们意识到收集数据代表着接受物联网，这意味着必须迅速掌握一项新兴技术。

____

____

3. 智能驾驶汽车不仅速度快、效率高，而且行动灵活，如果可能的话，它以后还会更舒适。

____

____

4. 人工智能和物联网是未来有很大发展空间和广阔前景的技术之一。

____

____

5. 由于物联网以云为基础架构，它天生就比人工智能更具可扩展性。

____

____

Translation at Paragraph Level

English to Chinese Translation

①Interconnected technology is now an inescapable reality—ordering our groceries, monitoring our cities, and sucking up vast amounts of data along the way. ②The promise is that it will benefit us all—but how can it? ③In San Francisco, a young engineer hopes to "optimize" his life through sensors that track his heart rate, respiration, and sleep cycle. ④In Copenhagen, a bus running two minutes behind schedule transmits its location and passenger count to the municipal traffic signal network, which extends the time of the green light at each of the next three intersections long enough for its driver to make up some time. ⑤What links these wildly different circumstances is a vision of connected devices now being sold to us as the Internet of Things. ⑥The technologist Mike Kuniavsky, a pioneer of this idea, characterizes it as a state of being in which "computation and data communication are embedded in and distributed through our entire environment…I prefer to see it for what it is: ideally, the colonization of everyday life by information processing".

本段一共六句话。段落翻译除了包含各种句子翻译的技巧之外，还会体现句子间的关系。例如，本段第二句中的 the promise 指的是第一句中描述的 an inescapable reality。因此在翻译第二句时应增译"这种前景""此种未来"类似表达。这两句可以译为："如今，互联技术已成为不可避免的现实——我们订购杂货、监控城市，并在此过程中获取大量数据。在这种前景下，所有人都将受益——但这怎么实现呢？"

在第三句和第四句中，翻译地点时应增加定语，以提供背景知识，如“美国旧金山”和“丹麦哥本哈根”。第四句中可以使用省略冠词和代词的译法。这两句可以译为：“在美国旧金山，一位年轻的工程师希望通过传感器来追踪心率、呼吸和睡眠周期，以‘优化’自己的生活。在丹麦哥本哈根，一辆晚点两分钟的公交车将位置和乘客人数传输给市政交通信号网络，该网络会延长接下来三个十字路口绿灯的时间，从而让公交司机补时。”

第五句中可以省略开头的连词 what，直接译为：“将这些截然不同的环境联系在一起的是将设备联网的设想，这些联网设备正以‘物联网’的形式出售给我们。”最后一句中的 in which 和 ideally 可以使用增译法，使句意更顺畅。整句可以译为：“技术专家迈克·库尼亚夫斯基是这一设想的先驱，他将其描述为一种状态。在这种状态中，‘计算和数据通信嵌入并分布在整个环境中……我更愿意看到它的本来面目：最理想的情况是通过信息处理掌控日常生活’。”

III. Translate the following paragraph into Chinese.

The Internet of Things isn't a single technology. About all that connects the various devices, services, vendors, and efforts involved is the end goal they serve: capturing data that can then be used to measure and control the world around us. Whenever a project has such imperial designs on our everyday lives, it is vital that we ask just what ideas underpin it and whose interests it serves. Although the Internet of Things retains a certain sprawling and formless quality, we can get a far more concrete sense of what it involves by looking at how it appears at each of three scales: that of our bodies (where the effort is referred to as the "quantified self"), our homes ("the smart home") and our public spaces ("the smart city"). Each of these examples illuminates a different aspect of the challenge presented to us by the Internet of Things, and each has something distinct to teach us.

____

____

____

____

____

____

Chinese to English Translation

①道德规则对人工智能至关重要。②阿西莫夫的“机器人第一定律”是“机器人不能伤害人类”，但是对于复杂的人工智能技术而言，这个定律会失效落空，因为在这种人工智能技术中，分析决策和道德决策之间的潜在分歧是一个大问题。③分析决策是仅

基于事件的可能结果做出决定，但因为分析决策不一定是社会可接受的决策，所以还需要道德决策推动人工智能做出选择。④最常见的例子是一辆自动驾驶汽车压过行人而不转向，因为转弯可能会导致其他行人和车内乘客受伤的风险更高。⑤因此，人工智能的运用应更加谨慎，避免误用和发生意外。

本段一共五句话。由于英汉句式结构的差异，段落与句子在汉译英时都会增译大量的冠词、连词、代词和助动词。此外，还应注意名词的单复数和主谓一致，如第一句中的 ethical rules 对应的谓语是 are。整句可以译为：Ethical rules are essential for AI."。第二句中的"机器人第一定律"应加冠词，并注意大小写；描述定律内容时可以使用 that 引导的表语从句；重复部分"在这种人工智能技术中"可以用非限定性定语从句 where 表示。第二句可以译为："Asimov's First Law of Robotics is that 'A robot may not injure a human being', but such law might fall short with complex AI technologies, where the potential diversion between analytical and ethical reasoning can become a major issue."。

第三句中"事件的可能结果"需要增译冠词和连词，"需要道德决策"是无主句，使用被动语态翻译时需要增加助动词 are。第三句可以译为："Analytical decisions are merely based on the likely outcome of an event, but ethical decisions are needed to drive decisions of artificial intelligence since analytical decisions do not necessarily match socially acceptable decisions."。第四句中定语多，需要添加连词 which 组成定语从句。整句可以译为："The most frequent example is the one of a self-driving car which decides to run over some pedestrians without turning since a potential turn would lead to a higher risk of injuring both other pedestrians and the passengers inside the vehicle."。第五句也可以使用被动语态进行翻译："Therefore, AI should be implemented with care and consideration to avoid misuse and unintended consequences."。

IV. Translate the following paragraph into English.

人工智能将在 2019 年快速增长，监管机构、警察和司法当局需要创造适当的环境以确保正确使用此技术。艾萨克·阿西莫夫的"机器人三定律"代表了规范人工智能的首次尝试。当前规范人工智能的尝试似乎更像是确定良好行为的一般原则，而不是实际法规，并且缺乏约束力和强制措施。规范人工智能必须进行国际合作，并在大型 IT 公司中建立道德监控部门。阿西莫夫的一句名言是："目前生活中最可悲的是，科学汇集知识的速度比社会汇集智慧的速度快。"

____

____

____

____

____

____

I. Choose one of the IoT applications mentioned in Text B in this unit and do research on how it is developed. Then write an outline of your research results and make a five-minute oral presentation to the class.

Name: ____________________
Definition: ____________________
Development: ____________________
Example(s): ____________________
Prospects: ____________________

II. Read the following case and answer the questions.

1. What benefits does smart parking bring to the drivers and the municipal managers in Montpellier?

2. What lessons can we learn from the case of Montpellier?

Smart Parking in Montpellier

What Is Smart Parking?

The transportation industry, from automated driving to traffic congestion management, is poised to see major benefits from the use of IoT solutions. One major use case is smart parking, which can get rid of greenhouse-gas-emitting, and the trouble of circling the block or parking deck to look for an open space. Instead, sensor-equipped parking spots connected to a mobile app can guide drivers to the nearest available space. Integration of a payment system could also streamline the process of fumbling around for coins and watching the clock to go feed a meter.

Parking in Montpellier

Montpellier is a small French city of Occitanie. Its 270,000 inhabitants suffer, as in every other city, from traffic jams and lack of available parking slots. Montpellier shopping center has limited parking slots and there are traffic jams in some streets habitually.

Ten-minute searching for car parking three times daily means more than 180 hours per year. Driving around to look for an available car slot wastes fuel, produces anxiety, and increases pollution and traffic congestion in city centers.

The Deployment of Smart Parking in Montpellier

The Montpellier Mediterranean Metropolis (MMM), a public project dedicated to implementing IoT projects, worked with Synox and Libelium to deploy a smart parking solution. The main goal of this project was to make traffic more fluid and increase rotation at parking spaces near the town hall and a shopping center.

Twenty Smart Parking nodes were installed by Synox in two different areas of the city: six at the surroundings of the Montpellier Town Hall and the rest on the nearby Parc Marianne district. These devices were installed in the surface of the roadway on the parking areas for people with reduced mobility and delivery services with the aim to relieve congestion, streamline traffic, and improve access to car parking areas. Besides, sensors also gather data about the temperature of the roadway which will be used by the metropolis road authorities to take action in case of the presence of ice sheets.

The smart parking sensor network is connected to the Metropolis LoRaWAN private network which has been deployed and provided by Synox, who has built a specific infrastructure to allow the entity to keep sovereignty in the data from end to end. Data management displays show this real-time information and indicators of the use of car parking slots to the apps of the drivers' mobile phones and the LED directional signage, which are strategically placed around the car park areas.

"Connected Parking" as part of the "Open Data" approach helps citizens to have more information and online data available. Users, laboratories, and other start-ups can use this data and benefit from it, creating new services for a smarter, more dynamic, and more environmentally-friendly city. "The Smart Parking project made it possible to easily deploy—on the LoRaWAN private network infrastructure—an analysis of the strategic car parking availability, for the city mobility services, the City Police, and soon the citizens themselves", explains Jérôme Fenwick, CTO at Synox.

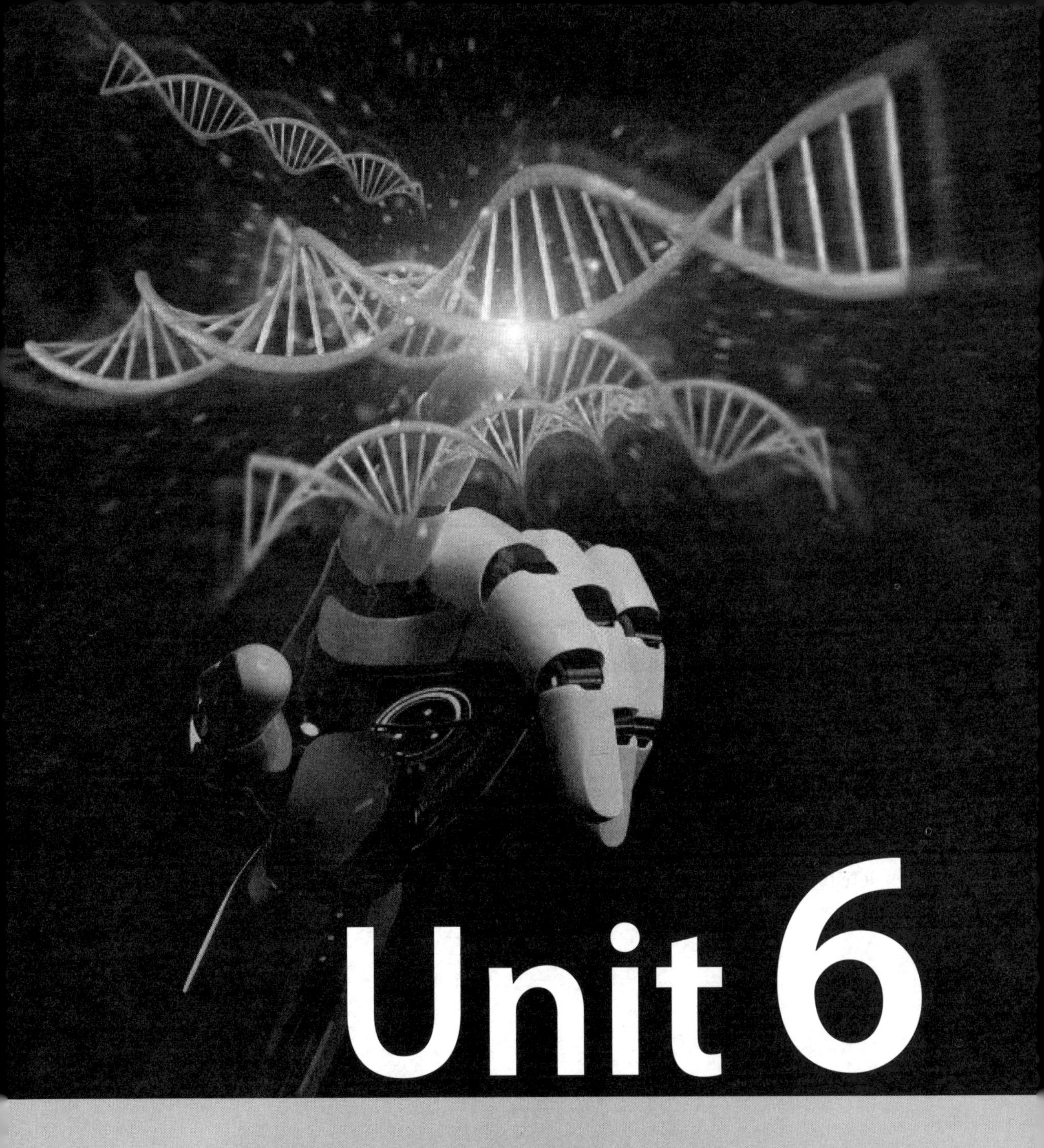

Unit 6

The Applications of Artificial Intelligence (I)

Learning Objectives

In this unit, you will learn:

- top ten industrial applications of artificial intelligence;
- the specific application of artificial intelligence in journalism;
- five roles of artificial intelligence in education;
- writing skills—the usage of exemplification;
- translating skills—translation of sentences without subjects.

Lead-in

I. Discuss the following questions with your partners.

1. Could you list some applications of artificial intelligence in our daily life?
2. What are the functions of these applications?
3. What benefits could these applications bring to us?

II. Work in pairs to list the companies engaged in the applications of artificial intelligence and their products.

No.	Companies	Products
1.		
2.		
3.		

Text A

Top Ten Industrial Applications of Artificial Intelligence

1 A long time ago, machines and computers were invented to decrease manual efforts and time to perform tasks efficiently. After a lot of growth and development in the field, people gradually get used to **utilizing** the technology. However, this is actually just a start because the forthcoming progress will make a greater shape in the form of artificial intelligence. Today almost every field has been affected by AI, though it is a branch of computer science. Tech giants like Google, Microsoft, and IBM are highly involved in researching and developing the technology which has already started bringing a revolutionary breakthrough. Although it is going to reform our future, we need to know how it's influencing our present life. So, in order to give you a glimpse of it, we are here with a list of top ten industrial applications of AI.

Journalism

2 In the technological era, people are used to reading blogs and articles online, but hardly do people realize that in fact most articles are produced by machines. Although it can't be used for writing in-depth articles, the simple reports that don't require much analysis can be easily prepared by AI. Companies like AP and Yahoo are using AI to prepare simple reports related to sports and elections that would take a lot of time if done manually. According to Automated Insights which introduced Wordsmith, many data-driven **entities** including real estate and e-commerce are using this platform. Based on current developing trend of machine writing, we could predict and witness that artificial intelligence will replace human writers to produce articles in different styles.

Entertainment

3 Artificial intelligence is widely used in the entertainment industry. Whether they're the video games or music apps, we all are well aware of the concept. Talking about games, the idea is not new and is being utilized from the very beginning, but today it has just increased exponentially. Games like Middle Earth and Far Cry are known for imparting personalities to the characters where they find objects, shoot, take cover, and do everything possible for victory. Similarly, for music and movies, the users are recommended to watch or listen to certain creations according to the choices

made in the past. If the user is frequently clicking on a specific genre, then that will be regarded as the section that he may take an interest in and it will be displayed in a separate section, saying "you may also be interested in".

Online Retail Stores

4 With the appearance and rising of online retail stores, people have developed the online purchase habit, which is quite at its peak right now. These websites also use artificial intelligence in certain ways like recommending the goods to the customers based on their past purchases or items put in the search box. Another way is offering chatbots for seeking guidance or solving queries.

Automobiles

5 It's a well-established fact that Google's driverless cars and Tesla's Autopilot features have already paved their way towards the introduction of AI in automobiles. Whether it's self-parking, detecting collision, blind spot monitoring, voice recognition, or **navigation**, it's almost like the car is playing the role as an assistant to the owner and teaching different ways of a safe driving.

Banking and Finance

6 As the volume of data increases, a large number of financial services have resorted to artificial intelligence. Robots are much quicker in analyzing market data to **forecast** changes in stock trends and manage finance as compared with the human **counterparts**. They can even use algorithms to provide suggestions for the clients concerning simple problems. In the similar way, banks are using AI to keep a track of the customer base, **addressing** their needs and suggesting different **schemes**. Often when there is a **suspicious** transaction from the users' **account**, they immediately get a mail to confirm that it's not an outsider who is carrying out that particular transaction.

Healthcare

7 In the healthcare industry, artificial intelligence is also carrying out a great deal of critical work. It is being used by doctors in assisting with the **diagnosis** and treatment **procedures**. This reduces the need for multiple machines and equipment, which in turn brings down the cost. IBM has invented Watson which is an artificial intelligent application. It has been designed to suggest different kinds of treatments on the basis of the patients' medical history and has been proved to be very effective.

Manufacturing

8 Manufacturing is one of the first industries that has been using AI from the very beginning. Robotic parts are used in the factories to assemble different parts and then pack them without any manual help. Right from the raw materials to the shipped final products, robotic parts play an **imminent** role in most of the entities.

Online Customer Service

9 Several websites offer customers to chat with their sales representatives in case of a query or **grievance**. However, most of the time these are not humans; rather these are chatbots that are trained to respond and extract the required knowledge from the site and present it to the customer. These bots utilize natural language processing to interpret the customer's query by focusing on the keywords and then in response, the required data is fetched. However, it's not quite easy for a robot to understand it because there is a huge difference between the human and machine language. Reportedly, progress is being made on that front too and hopefully, soon we'll be having a robot that could communicate with humans on its own.

Home Appliances

10 IoT technology also employ artificial intelligence including all the smart devices and **gadgets** used in our daily lives. The technique is to learn the behavior and usage pattern shown by the user and then **accordingly** the **appliance** starts behaving in a similar way on its own without any **instructions**. When it comes to a specific electronic appliance, then **thermostat** features make use of AI in an interesting way. It can set the temperature of your home just the way you want it at different hours of the day.

Smart Phones

11 Virtual personal assistant is the most common application of AI in our mobile phones. Siri, Cortana, and Google Now are some very commonly-used digital assistants that are found in iOS, Android, and Windows phones. These can answer whatever you ask them like "What movie is going to be released this Friday?" or "Who is Steve Jobs?" These assistants respond to the questions asked by the user through gathering information and then the required data is fetched to suit the user's preferences.

12 Artificial intelligence has been growing in all the technologically relevant fields, but it is also spreading in the areas where nobody has imagined it to be. This seems like an advance, but it can be equally **disruptive** in the future. It is believed that AI is a very sensitive issue and if not handled with care, it could end up imparting "superintelligence" to machines which would make them even more intelligent than humans. So, the bottom line is that it's great to have a machine by our side which can **imitate** human intelligence, but at the same time we have to think to what extent we are letting that thing take over.

Notes

WordSmith a software package primarily for linguists, in particular for work in the field of corpus linguistics. It is a collection of modules for searching patterns in a language. The software handles many languages.

Siri a virtual assistant part of Apple Inc.'s iOS, watchOS, macOS, HomePod, and tvOS operating systems. The assistant uses voice queries and a natural-language user interface to answer questions, make recommendations, and perform actions by delegating requests to a set of Internet services. The software adapts to users' individual language usages, searches, and preferences with continuing use. Returned results are individualized.

Google Now a feature of Google Search that offers predictive cards with information and daily updates in the Google app for Android and iOS. It proactively delivered information to users to predict (based on search habits and other factors) information they may need.

superintelligence an advanced form of artificial intelligence which possesses intelligence far surpassing that of the brightest and most gifted human minds. "Superintelligence" may also refer to a property of problem-solving systems (e.g., superintelligent language translators or engineering assistants).

Words and Expressions

utilize	/ˈjuːtəlaɪz/	*v.*	to use something, especially for a practical purpose 利用，运用，应用
entity	/ˈentəti/	*n.*	an organization (such as a business or governmental unit) that has an identity separate from other things 实体；存在
navigation	/ˌnævɪˈgeɪʃn/	*n.*	the skill or the process of planning a route for a ship or other vehicle and taking it there 导航；航行
forecast	/ˈfɔːkaːst/	*v.*	to indicate or estimate the future event or trend 预测，预报
counterpart	/ˈkaʊntəpaːt/	*n.*	a person or thing having the same function or position as somebody/something else in a different place or situation 对应的人或物
address	/əˈdres/	*v.*	to deal with 处理，对付
scheme	/skiːm/	*n.*	a plan or program of action 计划，方案
suspicious	/səˈspɪʃəs/	*adj.*	making you feel that something is wrong, illegal, or dishonest 可疑的，令人怀疑的
account	/əˈkaʊnt/	*n.*	a record of debit and credit entries 账户；账目
diagnosis	/ˌdaɪəgˈnəʊsɪs/	*n.*	the act of identifying a disease from its signs and symptoms 诊断
procedure	/prəˈsiːdʒə(r)/	*n.*	a medical operation 手术

imminent	/'ɪmɪnənt/	*adj.*	ready to take place 即将发生的，迫近的
grievance	/'griːvəns/	*n.*	something that you think is unfair and that you complain or protest about; a feeling that you have been badly treated 不满；投诉
gadget	/'gædʒɪt/	*n.*	a small tool or device that does something useful 小工具；小玩意
accordingly	/ə'kɔːdɪŋli/	*adv.*	for that reason 因此；于是
appliance	/ə'plaɪəns/	*n.*	an instrument or device designed for a particular use or function in the home 器具；电器
instruction	/ɪn'strʌkʃn/	*n.*	a piece of information that tells a computer to perform a particular operation 指令，指示
thermostat	/'θɜːməstæt/	*n.*	an automatic device for regulating temperature 恒温器
disruptive	/dɪs'rʌptɪv/	*adj.*	new and original, in a way that causes major changes to how something is done 颠覆性的
imitate	/'ɪmɪteɪt/	*v.*	to follow somebody/something as a pattern, model, or example 模仿，效仿

Exercises

Comprehension Check

I. Answer the following questions according to the text.

1. In what ways has AI been applied in the video games nowadays?
2. What services can AI offer in automobiles?
3. Why did banking and finance resort to AI?
4. How is AI applied in the home appliances?
5. What kind of convenience does AI provide for people in the use of smart phones?

II. Read the following statements carefully and decide whether they are true (T) or false (F) without turning back to check the text.

1. ______________ In the digital era, most articles online are actually created by robots.
2. ______________ Artificial intelligence can recommend music and movies to users according to their previous choices.

3. ______________ The current self-driving cars act as drivers to replace humans.

4. ______________ Artificial intelligence can help doctors to diagnose illness through analyzing the patients' medical history.

5. ______________ Chatbots are playing a vital role in online customer service and fully understand the customer needs.

III. Choose the best answer to each of the following questions according to the text.

1. Which is NOT mentioned as the application of AI in online retail stores?

   A. Providing payment services.

   B. Offering seeking guidance.

   C. Resolving doubts.

   D. Providing buying advice.

2. At present, banks are making use of AI to deal with various businesses EXCEPT ______________.

   A. meeting the clients' requirements

   B. tracing the clients' accounts

   C. warning the clients about financial risks

   D. offering insightful suggestions to the clients

3. How is artificial intelligence applied in the healthcare industry?

   A. AI replaces doctors to do operations independently.

   B. AI assists doctors in diagnosis and treatment process.

   C. AI plays the role as a nurse to attend patients.

   D. AI monitors the patients' health conditions and report anytime.

4. Artificial intelligence can be applied in the following fields EXCEPT ______________.

   A. healthcare　　B. smart phones

   C. home appliances　　D. cyber defense

5. What is the author's attitude towards the future applications of AI?

   A. Positive.　　B. Negative.

   C. Neutral.　　D. Indifferent.

Vocabulary Building

IV. Fill in the following blanks with the words and phrase given in the box. Change the form if necessary.

imminent	extract	imitate	utilize	resort to
accordingly	disruptive	witness	suspicious	diagnosis

1. ______________, the information may not accurately describe or reflect the software product when first commercially released.
2. Wimbledon 2018 has ______________ extensive use of artificial intelligence and Watson technologies to drive fan engagement and business efficiency.
3. AI could ______________ the specific data from the mass information instantly.
4. New software goes beyond passive recording to alerting law enforcement about ______________ activity in real time.
5. An exact ______________ can only be made by obtaining a blood sample.
6. Some surveyed charities said they were in ______________ danger of closing down due to financial reasons.
7. Someday, AI could learn to ______________ the existing knowledge to create new inventions.
8. But by lowering barriers to entry, ______________ technologies create expanded opportunities for experimentation and innovation.
9. No computer can ______________ the complex functions of the human brain.
10. In the near future, enterprises can ______________ AI for more business opportunities.

V. Match the words in the left column with the explanations in the right column.

1. entity	A. to go to where something is and bring it back
2. grievance	B. an organization (such as a business or governmental unit) that has an identity separate from other things
3. gadget	C. an exchange or transfer of goods, services, or funds
4. transaction	D. a small tool or device that does something useful
5. fetch	E. something that you think is unfair and that you complain or protest about; a feeling that you have been badly treated

Word Formation

VI. Fill in the following blanks with the words in capitals. Change the form if necessary. An example has been given.

e.g. *This discovery has* revolutionized *our understanding of how the human brain works.* **REVOLUTION**

1. The country recently announced plans to begin ______________ its army. **MODERN**

2. As Chinese companies push to ______________ their operations, they're keen to hire more foreign students who speak languages other than Chinese. **GLOBAL**

3. The government hopes to ______________ some of the agricultural regions. **INDUSTRY**

4. We encourage users to change their passwords on a regular basis and also ______________ themselves with our online safety tips at security. **FAMILIAR**

5 I find it difficult to ______________ with him when he complains so much. **SYMPATHY**

Translation

VII. Translate the following sentences into English with the words and phrases in brackets.

1. 根据目前机器写作的发展趋势，我们可以预测并见证人工智能将取代作家写出不同风格的文章。(base on; predict; replace)

____

____

2. 同样地，根据用户过去的选择记录，人工智能可以为其推荐某些音乐和电影作品。(similarly; recommend; creation)

____

____

3. 这些应用程序通过收集信息来回应用户提出的问题并获取所需的数据，以满足用户的偏好。(respond to; fetch; preference)

____

____

The Washington Post Has Published 850 Articles in the Past Year

1 It's been a year since *The Washington Post* (short for *The Post*) started using its **home-grown** artificial intelligence technology, Heliograf, to produce about 300 short reports on the Rio Olympics. Since then, it has made use of Heliograf to **cover congressional** and **gubernatorial** races on Election Day and D.C.-area high school football games. The Associated Press (AP) has used robots to automate earnings coverage, while *U.S.A. Today* has made use of video software to create short videos. But media **executives** are more excited about AI's potential to go beyond rote reporting. Jeremy Gilbert, director of strategic initiatives at *The Post*, shared what the paper has learned so far from robo reporting and what it's still trying to figure out.

Robo Reporting Can Expand Articles

2 *The Post* has used Heliograf to create around 850 articles in its first year. That included 500 articles around the election that generated more than 500,000 clicks—not a ton in the scheme of things, but *The Post* wouldn't plan to put a lot of manpower in most of these stories anyway. For the 2012 election, for example, *The Post* did just 15% of what it generated in 2016.

Robots Can Help Journalists

3 Media outlets say that adopting AI could enable journalists to do more high-value work instead of taking their jobs away. The AP predicts that it's freed up 20% of reporters' time spent covering **corporate** earnings and that AI continues improving its **accuracy**. "Regarding the automated financial news coverage by AP, the error rate in the copy decreased even as the amount of the output increased more than **tenfold**," said Francesco Marconi, AP's strategy manager and AI co-leader.

4 *The Post* is also trying to find out the effective way to help its journalists with **substantive** reporting by using Heliograf. During the election, it utilized Heliograf to inform the newsroom when election results started to change in an unexpected direction, giving reporters lead time to fully cover the news. Gilbert wants Heliograf to play a more significant role in the following election. He also sees the potential for Heliograf to do **legwork** for reporters in other ways, like

spotting trends in financial and other big data sets. "We think we can help people find interesting stories," he said. Heliograf also can be deployed to update ongoing stories like weather events in real time, providing a service for readers.

AI Has B2B Applications

5 The traditional business model of journalism is the ad-supported and stressed pageview model. Publishers need to get readers or other groups to pay to support their business models. "Right now, automated journalism is about producing volume. **Ultimately**, media companies will have to figure out how to go beyond the pageview," said Seth Lewis, a journalism professor at the University of Oregon whose focuses involve the rise of AI in media. *The Post* has had conversations about what AI could do, which has a business-to-business application, but hasn't taken the idea further, Gilbert said. "It has a widespread utility that goes beyond individual news consumers. The target can also be people who are interested in very specific things."

Jury's Out on News Impact and the Economic Benefits

6 Robo reporting can serve a lot of **niche** audiences that, added up, can increase a news outlet's reach in a meaningful way. That's the thinking behind the local football coverage. It's unclear how that approach can be scaled to cover local communities, where the digital news model has fallen short. Heliograf can be used to **digest** data like standardized test scores and crime statistics; covering a zoning board meeting is another matter. And AI isn't being used beyond big news organizations, Lewis pointed out. "There's such a big difference between the AI haves and have-nots. There is a long way for us to witness these things being **implemented** at the local level."

7 When it comes to the economic benefits, it's not easy to value profits generated by the robo reporting. Right now, *The Post* can count the stories and pageviews that Heliograf produced. It's even harder to **quantify** its impact on how much time it gives reporters to do other work and the value of that work. It's also hard to quantify how much engagement, ad revenue, and **subscriptions** can be attributed to those robo-reported stories. (On the resource side, now that it's built, Heliograf has about five people dedicated to it, not including editors that it borrows to help figure out how to apply it.) "We're still starting to figure out what the economic impact is going to be when it makes sense to automate," Gilbert said.

Notes

The Washington Post a major American daily newspaper published in Washington, D.C., with a particular emphasis on national politics and the federal government. It has the largest circulation in the Washington metropolitan area.

Heliograf part of a suite of artificial intelligence tools which will be available through *The Washington Post*'s publishing platform Arc. It is an in-house program that automatically generates short reports for their live blog.

Rio Olympics the 2016 Summer Olympics, commonly known as Rio 2016, an international multi-sport event that was held from August 5 to 21, 2016 in Rio de Janeiro, Brazil. These were the first Olympic Games ever to be held in South America and the fourth to be held in a developing country.

Election Day the day set by law for the general elections of federal public officials in the United States. It is statutorily set as "the Tuesday next after the first Monday in the month of November" or "the first Tuesday after November 1".

The Associated Press a multimedia news agency headquartered in New York City. It has been a leader in news innovation with a commitment to producing world-class journalism and advancing the power of facts since 1846.

B2B business-to-business (in some countries, B to B), a situation where one business makes a commercial transaction with another.

Words and Expressions

home-grown /ˌhəʊm 'grəʊn/ *adj.* made, trained, or educated in your own country, town, etc. 自主研发的；国产的

cover /'kʌvə(r)/ *v.* to report news about an event 采访，报道

congressional /kən'greʃənl/ *adj.* related to or belonging to a congress or the Congress in the U.S. 国会的；议会的

gubernatorial /ˌguːbənə'tɔːriəl/ *adj.* of or relating to the state governor 州长的，州长职位的

executive /ɪg'zekjətɪv/ *n.* a person who has an important job as a manager of a company or an organization 主管领导，管理人员

corporate /'kɔːpərət/ *adj.* related to a corporation 公司的，企业的

accuracy /'ækjʊrəsi/ *n.* the state of being exact or correct; the ability to do something with skills and without making mistakes 精确度

tenfold /'tenfəʊld/ *adv.* by or up to 10 times as many or as much 十倍地，成十倍

substantive /'sʌbstəntɪv/ *adj.* dealing with real, important, or serious matters 实质性的；重大的

legwork /'legwɜːk/ *n.* difficult or boring work that takes a lot of time and

			effort, but that is thought to be less important 跑腿活儿
ultimately	/ˈʌltɪmətli/	*adv.*	in the end; finally 最后，最终
niche	/niːʃ/	*n.*	a small section of the market for a particular service or product 细分市场
digest	/daɪˈdʒest/	*v.*	to obtain an idea or meaning from something 领会；消化；吸收
implement	/ˈɪmplɪment/	*v.*	to carry out; to accomplish 实施，执行
quantify	/ˈkwɒntɪfaɪ/	*v.*	to describe, or measure something as an amount or a number 量化，定量
subscription	/səbˈskrɪpʃn/	*n.*	an amount of money you pay, usually once a year to receive a service; the act of paying this money 订阅，订购

Exercises

Comprehension Check

I. Identify the paragraph from which the information is derived.

1. ____________ Robo reporting could increase the number of articles.
2. ____________ Media companies tried to get rid of the pageview model to find out a new operation pattern.
3. ____________ *The Washington Post* has applied AI technology in writing reports related to sports and elections.
4. ____________ Robo reporting could help expand the niche market.
5. ____________ AI has a big potential to cover local communities.
6. ____________ It's tough to measure the economic impact of Heliograph.
7. ____________ Heliograf is expected to do more information collection work.
8. ____________ AI shows high accuracy in covering financial news.
9. ____________ Robots could save the reporters' time to do more creative work.
10. ____________ Now, *The Washington Post* evaluates the value of robo reporting through the number of articles it has produced.

II. Answer the following questions according to the text.

1. In what ways does AI help the journalists?
2. What is the traditional profit model of journalism?
3. How are the economic benefits of robo reporting evaluated?

Vocabulary Building

III. Fill in the following blanks with the words and phrase given in the box. Change the form if necessary.

homegrown	coverage	rote	accuracy	corporate
substantive	legwork	standardize	implement...into	quantify

1. China has made no secret of its desire to develop a strong, ______________ auto industry.
2. He received a national news Emmy Award for his ______________ of the 1988 Sudan famine.
3. The problem starts with an educational system that stresses ______________ learning instead of creativity.
4. We have to balance the need for ______________ with the need to provide timely information.
5. With plenty of weak companies at risk, unforeseen ______________ bankruptcies could upset the government's orderliness.
6. The talks were the first ______________ negotiations between the two sides since the war.
7. The popularity of social networks like LinkedIn makes this kind of ______________ easier than ever.
8. In order to ______________ work and improve productivity, they rely on automated procedures and laboratory robots.
9. The greatest strength, but also the greatest weakness of high tech, is that whenever a cool new technology comes along we all race to ______________ it ______________ whatever we are doing.
10. Economists use this measurement when they are trying to ______________ the amount of money in circulation.

Text C

Five Roles of Artificial Intelligence in Education

1 For decades, science fiction authors, futurists, and movie makers have been predicting the fantastic and sometimes devastating changes that will occur with the advent of widespread artificial intelligence. So far, AI hasn't made any such crazy waves, and in many ways it has quietly become ubiquitous in many aspects of our daily lives. From the intelligent sensors that help us take perfect pictures to the automatic parking features in cars, artificial intelligence of one kind or another is all around us, all the time. Education is one field that artificial intelligence is ready to make huge changes.

2 While **humanoid** robots may not play the same role as teachers within the next decade, there are many projects already in the works that use computer intelligence to help students and teachers gain the educational experience. Here are just a few ways those tools will shape and define the educational experience of the future.

adj. 类人的

Educational Software Can Be Tailored for Student Needs

3 One of the key ways that artificial intelligence will make a great influence on education is through the application of greater levels of individualized learning. Some of this is already happening through growing numbers of **adaptive** learning programs, games, and software. These systems are designed to meet the needs of students, putting greater emphasis on certain topics, repeating things that students haven't grasped, and generally helping students study at their own pace, whatever that may be.

adj. 适应的

4 This kind of custom-**tailored** education could be a machine-assisted solution to helping students at different levels work together in

adj. 特制的

one classroom, with teachers facilitating the learning and offering help and support when needed. Adaptive learning has already exerted a dramatic impact on education (especially through programs like Khan Academy), and as AI keeps evolving in the coming decades, adaptive programs like these will possibly get enhanced and expand constantly.

AI-Driven Programs Can Give Students and Educators Helpful Feedback

5 AI can not only help teachers and students to **craft** courses that are adapted to their needs, but also provide feedback for both about the success of the course as a whole. Some schools, especially those offering online courses and learning programs, are making use of AI systems to **monitor** student progress and to alert professors when there might be an issue with student performance.

v. 精心制作

v. 监测；跟踪

6 These kinds of AI systems allow students to get the support they need and professors to find areas where they can improve instruction for students who may struggle with the subject matter. However, AI programs at these schools aren't just providing suggestions on individual courses. Some are dedicating to inventing systems which can help students select majors on the basis of areas where they succeed and struggle. While students don't have to adopt the advice, it could **engender** a new means of college major selection for future students.

v. 产生

AI Can Make Trial-and-Error Learning Less Intimidating

7 Trial and error is a vital part of learning, but for many students, the idea of failing, or even not knowing the answer, is frustrating and depressing. Some would feel anxious and embarrassed in front of their peers and teachers. An intelligent computer system, designed to help students to learn, is a much less **daunting** way to deal with trial and error. Artificial intelligence could offer students a way to experiment and learn in a relatively judgment-free environment, especially when AI tutors can offer solutions for improvement. In fact, AI is the perfect format for supporting this kind of learning, as AI systems themselves often learn by a trial-and-error method.

adj. 使人气馁的

AI Could Change the Role of Teachers

8 Teachers have been playing an indispensable role in education for

centuries but what that sort of role is and what it entails may change due to new technology in the form of intelligent computing systems. AI can take over tasks like grading, can help students improve learning, and may even be a **substitute** for real-world tutoring. Furthermore, AI could be adapted to many other aspects of teaching as well. AI systems could be designed to provide expertise, serving as a place for students to ask questions and find information, or could even potentially take the place of teachers for very basic course materials. In most cases, however, AI will shift the role of teachers to that of facilitators.

n. 代替者；代替物

9 Teachers will **supplement** AI lessons, assist students who are struggling, and provide human interaction and hands-on experiences for students. In many ways, technology is already driving some of these changes in the classroom, especially in schools that are online or embrace the **flipped** classroom model.

v. 补充

adj. 翻转的

Data Powered by AI Can Change How Schools Find, Teach, and Support Students

10 Intelligent computer systems could realize smart data gathering, which is already making changes about how colleges interact with **prospective** and current students. From recruiting to helping students choose the best courses, intelligent computer systems are helping make every part of the college experience more closely tailored to student needs and goals.

adj. 未来的，预期的

11 In today's higher-ed area, data mining systems are already playing an integral role, but artificial intelligence could alter higher education to a greater extent. Some schools have taken the initiatives to offer students AI-guided training which can ease the transition between high school and college. Who knows but that the college selection process may end up a lot like Amazon or Netflix, with a system that recommends the best schools and programs for students' interests?

12 The result? Education could look a whole lot different a few decades from now.

Exercises

Comprehension Check

I. Answer the following questions according to the text.

1. What changes has AI made in our daily life?
2. How does educational software meet students' learning needs?
3. What kind of role does AI play in some schools which provide online courses?
4. In what ways will AI make students feel less anxious and awful in learning?
5. What sort of work could AI take over from teachers in the future?

II. Read the following statements carefully and decide whether they are true (T) or false (F) without turning back to check the text.

1. ______________ Artificial intelligence has made a great influence on education by the application of personalized learning.
2. ______________ AI programs could help students select courses and majors based on their interests.
3. ______________ Trial and error is indispensable in the learning progress that many students could deal with it readily.
4. ______________ AI will change the role of teachers from an educator to a developer.
5. ______________ In the future, AI systems could offer advice on the most suitable colleges for high school students based on their interests.

III. Discuss the following questions based on your understanding of the applications of artificial intelligence.

1. Can you imagine what role teachers will play in the future with the widespread use of AI in education?
2. How will AI technology change the way teachers interact with students?
3. Can you list some other functions that AI could perform in education?

Writing Skills

例证法

例证法（exemplification）是通过具体、生动的例子来解释、说服、定义或阐明观点。它是最常见也是最有效的段落展开方法之一，它可以轻松地将主题句扩展成一个具有说服力的篇章。科技类的文章大多需要使用例证法来说明和阐释观点。使用这种写作方法时，首先需要摆明论点，再列举一系列的论据对其进行陈述和解释，或用具体的事例进行说明。本单元的 Text A 列举了大量生活中的实例来说明人工智能的应用，使读者很容易就理解了人工智能是如何应用于各个行业的。

例证法常用的连接词有：for example, as an example, as a case in point, such as, etc.; first and foremost, initially, to begin with, in the first place, etc.; what's more, besides, furthermore, in addition (to/that), moreover, etc.; finally, eventually, last but not the least, etc. 等。

例 1：When it comes to AI applications, let's take Google as an example.

例 2：AI has been applied into various fields, such as manufacturing, journalism and entertainment.

那么，如何用例证法来展开一个段落呢？首先，段落一开始需要使用主题句来阐明将要讨论的主要观点或主题；其次，选择并列举具体和恰当的例子来说明和例证这一观点或主题；最后，总结这一段落。例如：

Topic sentence:

Despite the progress that many other industries have made, healthcare is likely to be the one market where AI can truly have an impact that goes beyond convenience and positively affects human lives.

Examples:

For example, AI can be a personal life coach to create a customized experience for each individual patient and offer proactive alerts that can be sent back to physicians.

Another example is healthcare bots. Bots can be used in situations such as scheduling follow-up appointments with a patient's provider online. They can also help a patient with his/her medication or medical billing needs.

Write a short paragraph about the applications of artificial intelligence in smart phones with specific examples.

____

____

____

____

____

____

无主句（sentences without subjects）是指句子中的谓语动词没有相应的逻辑主语。汉语和英语都有无主句，其原因却各不相同。汉语表达突出“以人为本”，通常以“人”作为谓语动词的逻辑主语。有时候主语是众所周知的某人或某事，则可以省略主语。英语中通常会用大量的抽象名词作主语，表示某种现象或理念，主语与谓语动词之间不是简单的“人—动作”的关系。因此，翻译时要充分考虑两种语言自身的规律以及语言背后的文化差异。常见的无主句翻译方法包括以下四种：

1. 根据语境添加必要的主语

例 1：在摄影、人工智能、通信、设计等领域取得了多项突破性创新。

We have achieved many breakthroughs in domains such as photography, artificial intelligence, communications capabilities, and design.

2. 表示客观存在的句子，选择 there be 句型

例 2：在许多情况下，可以将对象分类为数字图像。

There are many situations where you can classify the object as a digital image.

3. 无灵主语（没有生命的物体作主语）时，选择被动语态

例 3：相比之下，当用于训练的信息既没有分类也没有标记时，则使用无监督机器学习算法。

In contrast, unsupervised machine learning algorithms are used when the information used to train is neither classified nor labeled.

4. 前一句的整句内容是后一句的逻辑主语，选择非限定性定语从句

例 4：然后我们将使用一个名为 NumPy 的 Python 模块进行一个非常快速的示例，可以通过它进行抽象的矩阵计算。

Then we'll go through a really quick example using a module for Python called NumPy, which allows you to abstract and matrix calculations.

Translation at Sentence Level

I. Translate the following sentences into Chinese.

1. Care should be taken to ensure that the pulse signals shall allow no interruptions during operation.

   ____

   ____

2. Make sure that the machine is level before starting it.

   ____

   ____

3. It is essential to lay out a plan for intelligent logistics to achieve transformation upgrading.

   ____

   ____

4. There will be over 200 drone airports built in Beijing to deliver the packages to any other city in China within 24 hours.

   ____

   ____

5. It is shown in a patent application that Amazon plans to customize product recommendations based on consumers' demands by drones.

   ____

   ____

II. **Translate the following sentences into English.**

1. 方便操作者自动规划最优路线，从而拿到相应订单的商品。

____

____

2. 通过收购一个顶级计算机团队，完成了针对云处理系统的升级。

____

____

3. 智能物流集多种服务功能于一体，体现了现代经济运作特点的需求。

____

____

4. 取快递时，只要打开手机 app 扫一扫就可以体验智能查件。

____

____

5. 在产品创新方面需要依靠人工智能不断提升产品性能，打造“更懂你”的智慧手机。

____

____

Translation at Paragraph Level

English to Chinese Translation

①Will we ever see robots put on our make-up for us? ②A number of gadgets released in the last few years suggest we might. ③Take the Opté wand from Proctor and Gamble (P&G), a make-up printer unveiled at this year's Consumer Electronics Show in Las Vegas. ④The wand scans the skin and precisely applies tiny amounts of make-up to hide age spots, burst blood vessels, and other blemishes. ⑤Its tiny built-in camera takes 200 frames per second, while a microprocessor analyses this data to differentiate between light and dark areas. ⑥A micro printer then applies the foundation to your skin. ⑦P&G, which hopes to launch the product by 2020, says the printer's precision means it needs relatively little serum, so people's make-up bills should drop. ⑧Imagining where the trend could go, design agency Seymour Powell has unveiled a printer concept that would allow make-up looks seen online to be downloaded and printed directly onto the face.

本段一共八句话，主要介绍了“人工智能将帮助人类化妆”这一话题，全段的句式不是很复杂，需要注意的是不同语境中词汇的翻译方法。

第一句中的 put on make-up 可译为“化妆”，ever 可译为“有一天”；第二句中的 gadgets released 可译为“推出的产品”。这两句可以译为：“有一天机器人会帮我们化妆吗？近几年推出的多个电子产品告诉我们，这有可能发生。”第三句中的 take 可译为“以……为例”，unveiled 可译为“亮相或发布”。这句话可以译为：“以宝洁公司研发的 Opté 魔杖为例，这款化妆品打印机今年在拉斯维加斯举行的消费类电子产品展览会上亮相。”第四句中的 applies…to 可译为“涂抹”，tiny amounts of 可译为“少量”，age spots 译为“色斑”，burst blood vessels 可译为“血丝”。这句话可以译为：“这款打印机可以扫描皮肤，精准地在皮肤上涂抹少量化妆品来遮盖色斑、血丝和其他瑕疵。”

第五句中 while 引导的从句表示“还”，built-in 可译为“内置的”，takes 200 frames 可译为“秒速达到 200 帧”，light and dark areas 在翻译时需要使用增词法，即加定语译为“皮肤的亮区和暗区”。这句话可以译为：“打印机内置的微型摄像机秒速可达到 200 帧，还有一个分析数据的微处理器，可以区分皮肤的亮区和暗区。”第六句中的 foundation 一词需要结合语境，可以译为“粉底液”。这句话可以译为：“微型打印机可以将粉底液涂抹到你的皮肤上。”

第七句中的 launch the product 可译为“推出这一产品”，make-up bills should drop 可译为“减少人们在化妆品上的开支”。这句话可以译为：“宝洁希望在 2020 年前推出这一产品，并表示打印机的精准度意味着需要的精华液相对更少，从而可以减少人们在化妆品上的开支。”第八句的前半句是现在分词引导的伴随状语，共用了后半句中的主语 design agency Seymour Powell，根据汉语的表达习惯，翻译时需要把主语提前。此外，unveil 可译为“提出”，make-up looks 可译为“妆容”。这句话可以译为：“关于这一趋势的未来，设计公司西摩·鲍威尔提出了一个打印机概念，即未来客户可以下载网上看见的妆容，并直接打印到自己的脸上。”

III. Translate the following paragraph into Chinese.

China's state-run news agency Xinhua debuted its first artificial intelligence news anchors at the ongoing Fifth World Internet Conference that kicked off in Wuzhen, east China's Zhejiang Province. In the video clips released by Xinhua, the AI news anchors resemble real people, as they were modelled on the news anchors working in the agency. They can deliver the news just like human anchors as their machine learning program can extract and synthesize the voice, lip movements, and facial expressions of real anchors, according to Xinhua. As editors input the news, the AI anchors tirelessly, quickly, and accurately report it throughout the day. According to Xinhua, the AI anchors have already joined the daily news reporting team and worked 24 hours a day reporting news on the agency's social media platforms, including the news app, official WeChat, and Weibo accounts as well as the TV Web page, bringing audiences "a brand-new news experience". Xinhua said that the AI anchors have immeasurable prospects for the future news reporting as they could reduce production costs and improve efficiency and accuracy.

____

____

____

____

____

____

Chinese to English Translation

①由微软亚洲研究院与雷德蒙德研究院的研究人员组成的团队近日宣布，其研发的机器翻译系统在通用新闻报道的中译英测试中达到了人类专业译者水平。②这是首个在新闻报道的翻译质量和准确率上可以比肩人工翻译的翻译系统。③微软的这次突破将机器翻译超越人类业余译者的时间提前了整整七年，远远超出了众多研究人员的预期。④虽然此次突破意义非凡，但微软研究人员也提醒大家，这并不代表人类已经完全解决了机器翻译的问题，只能说明人们离终极目标又更近了一步。⑤有了人工智能、云计算、智能语料数据库和正在研发的神经机器技术的加持，神经机器翻译将会成为今后机器翻译的绝对主流。

本段一共五句话。第一句中“团队”一词前的定语很长，汉语的表达习惯是把定语放在修饰的核心词之前，而英语的表达习惯则是突出核心词，再加定语进行说明。后半句中包含了一个条件从句“在通用新闻报道的中译英测试中”，在翻译时需要使用引导词，并且这句话的主语被省略了，变成了一个无主句，动词可以使用分词形式。第一句可以译为：“A team of researchers from the Microsoft Research Asia and Research Redmond announced that, when reporting and translating general news between Chinese and English, its machine translation system has reached the level of human professional translators.”。

第二句的主干部分是“这是首个翻译系统”，其余部分都是“翻译系统”的定语。在翻译时需要先译出主句，其他部分用定语从句来翻译。整句话可以译为：“This is the first ever system that could be on a par with human professionals in terms of quality and accuracy.”。**第三句和第二句类似，主干部分是“这次突破超出了众多研究人员的预期”，其他部分可用定语从句来翻译。在翻译时也需要先译出主句，再用定语从句翻译。整句话可以译为：**“Microsoft's breakthrough, which takes the machine translation beyond the time of human amateur translators, has advanced seven years ahead of the expectations of numerous researchers.”。

第四句是一个让步状语从句，可以使用 although **来引导句子。整句话可以译为：**

"Although the breakthrough is remarkable, Microsoft researchers also remind you that this doesn't mean humans have completely solved the problem of machine translation; we can only explain that we're closer to the goal."。第五句的主句是"神经机器翻译将会成为今后机器翻译的绝对主流"，前半句是主句的条件。整句话可以译为："Based on AI, cloud-computing, smart corpus database, and the ongoing development in neuromachine technology, neuromachine translation would be the mainstream media in insofar machine automation translation."。

IV. Translate the following paragraph into English.

埃隆·马斯克（Elon Musk）近日表示，其创业公司"神经连接"（Neuralink）即将公布首个脑机交互界面，该界面可将人脑和计算机连接起来。神经连接公司于 2016 年成立，其宏大的目标是开发计算机硬件来增强人脑功能，但具体实现方式鲜少对公众透露。马斯克已屡次重申，人工智能的飞速发展已经威胁到了人类的生存。他认为，如果将来人类要与智能科技竞争，那么这样的脑机交互非常重要。专家提醒人们，尽管这项技术有可能增强人脑功能，但是脑机交互有可能被变异的人工智能劫持。一旦被劫持，这意味着人工智能可以通过脑机连接控制人的思维、决定和情感。

____

____

____

____

____

____

Workshop

I. Choose one of the AI applications mentioned in Text A in this unit and do research on how it is developed. Then write an outline of your research results and make a five-minute oral presentation to the class.

Name: ____________________

Leading Company: ____________________

Achievements: ____________________

Developing Process: ____________________

Functions: ____________________

II. Read the following case and answer the questions.

1. What problem(s) should DigitalGenius deal with?
2. What sort of AI technology has been adopted by the company to solve the problem? And how?
3. What are the results of applying the AI technology in online education?

Online Education Company Improves Customer Support with Auto-Suggestion of Macros

Technology Provider: DigitalGenius is an artificial intelligence solutions provider for customer service operations.

User Company: Magoosh provides online educational tools for students preparing for standardized tests, such as GRE, GMAT, etc.

Problem

Magoosh's support staff comprises two teams of 50 agents: a community support team for handling account inquiries, and remote tutors to handle in-depth questions for specific tests. Magoosh uses Zendesk to handle its customer support requests. It has over 900 macros on Zendesk, which are pre-written, standard responses to common questions asked by the company's customers.

The support staff found it difficult to search or discover these macros for offering timely customer help, which they believed to be negatively affecting their customer satisfaction scores on responses to questions about standardized tests. Part of this searchability problem was the enormous number of macros which took a lot of time to search through and manage.

Actions Taken

The DigitalGenius AI platform was integrated with Magoosh's Zendesk console. DigitalGenius trained a deep neural network to analyze incoming customer inquiries based on historical customer logs—learning how Magoosh's support staff replied to various incoming inquiries.

DigitalGenius then automatically suggests the most relevant macros for new customer inquiries so the support team does not spend time searching for macros or manually composing new responses to common customer inquiries. DigitalGenius claims that its AI platform achieves this automatic macro suggestion by using deep learning models to extract the meaning and context of incoming inquiries and predicting the expected response. In addition, the platform has a historical response search feature, which the support staff can access.

When asked about this historical response search feature, Juan Ageitos of DigitalGenius told us:

> "The historical response search feature looked for historical tickets in which customers asked similar questions to the one agents were working on. We built the search ourselves, using our own search algorithms. And beyond that, we have a different UI to the Zendesk search, including the ability for historical response searches to take place in the app sidebar, so agents don't have to navigate away from the page.
>
> The coolest feature is that we prioritized historical tickets that had the highest CSAT. When agents searched from historical responses, we displayed to them whether that ticket got a high or low CSAT rating. So we think our feature promoted the best answers."

The platform also reportedly predicts the relevant metadata about the case, such as tags, inquiry type, priority, and other case details. With this information, it is able to analyze and route cases to the appropriate team. For example, if the incoming query is an account inquiry, the platform routes the request to the community support team, and for in-depth educational queries, it routes the requests to remote tutors—eliminating the need for a "human filter" to handle all tickets.

Results

According to DigitalGenius, about 83% of all customer tickets are supported by the DigitalGenius platform integrated with Magoosh's Zendesk. The company also claims a 92% accuracy in case tag predictions (tags are used within Zendesk for case categorization—for example, "refunds" might be a tag for that particular kind of customer issue). This improvement happened over an initial 6-month period with Magoosh, which Ageitos describes as a "learning segment"—stating that new and updated projects are underway with Magoosh now.

Asked for clarification on what it means to have 83% of messages "supported" by DigitalGenius, Ageitos replied:

> "Supported in this context means that DigitalGenius AI has assisted Magoosh with 83% of their tickets—whether this is classifying them, suggesting the right macro, or automating a response. The remaining 17% escaped our current AI capabilities, and had be dealt with manually in order to provide the best possible answer and avoid a potential wrong answer to the customer."

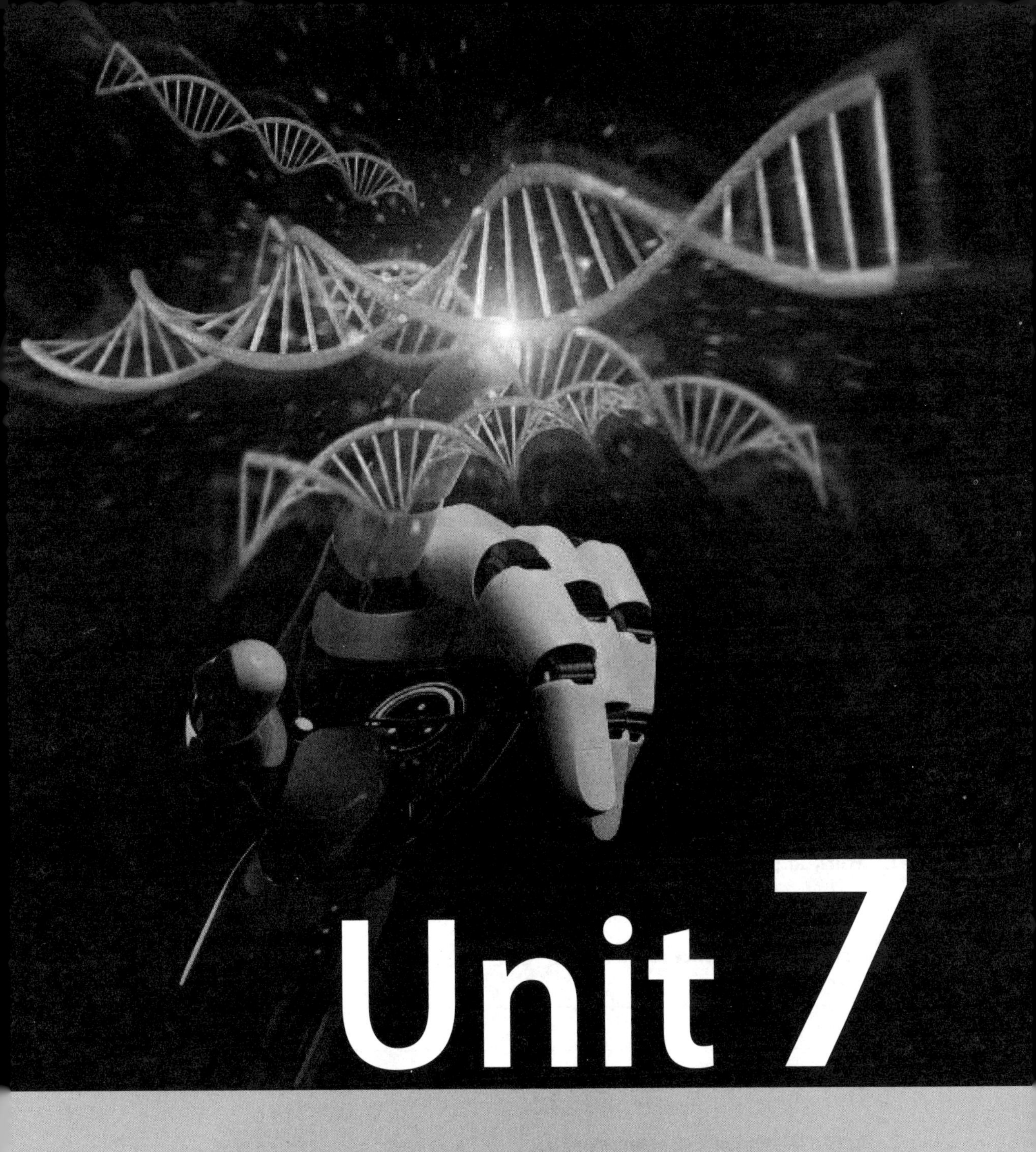

Unit 7

The Applications of Artificial Intelligence (II)

Learning Objectives

In this unit, you will learn:

- top ten applications of artificial intelligence in the field of healthcare;
- the applications of artificial intelligence in face recognition in China;
- high returns on drug discovery and development brought by artificial intelligence;
- writing skills—the usage of non-finite verbs;
- translating skills—splitting and combination.

Lead-in

I. Discuss the following questions with your partners.

1. Do you know anything about the applications of artificial intelligence in the field of healthcare?

2. Do you think there will be a large number of laid-off doctors in the future because of the application of artificial intelligence? Give your reasons.

II. Work in pairs to discuss whether artificial intelligence is a friend or an enemy to doctors. List reasons to support your idea.

Artificial intelligence as a friend:

1. ____________________
2. ____________________
3. ____________________

Artificial intelligence as an enemy:

1. ____________________
2. ____________________
3. ____________________

Text A

Top Ten Applications of Artificial Intelligence in Healthcare

1 There's a lot of excitement right now about how artificial intelligence is going to change healthcare. And many AI technologies are **cropping up** to help people **streamline** administrative and clinical healthcare processes. According to venture capital firm Rock Health, 121 health AI and machine learning companies raised $2.7 billion in 206 deals between 2011 and 2017.

2 The field of health AI is seemingly wide—covering wellness to diagnostics to operational technologies—but it is also narrow in that health AI applications typically perform just a single task. We investigated the value of ten promising AI applications and found that they could create up to $150 billion in annual savings for U.S. healthcare by 2026 (see Figure 7.1).

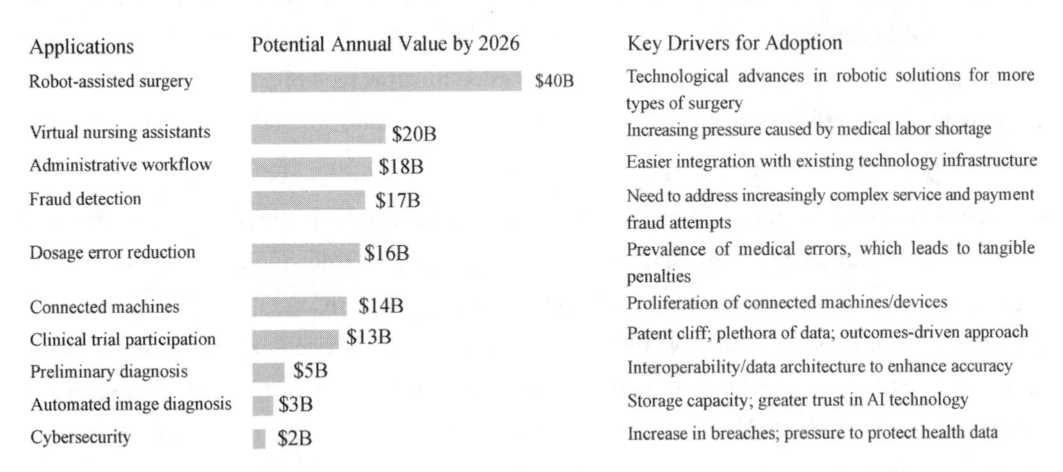

Figure 7.1 Ten AI applications that could change healthcare
(Source: Accenture)

3 We identified these specific AI applications based on how likely adoption was and what potential exists for annual savings. We found AI currently creates the most value in helping frontline **clinicians** be more productive and in making back-end processes more efficient, but not yet in making clinical decisions or improving clinical outcomes. Clinical applications are still rare.

4 Let's take a look at a few examples of AI on the frontline of healthcare. AI has demonstrated its

aptitude for improving the efficiency of image analysis by quickly and accurately flagging specific **anomalies** for a **radiologist**'s review. In 2011, researchers from NYU Langone Health found that this type of automated analysis could find and match specific lung **nodules** (on chest CT images) between 62% to 97% faster than a panel of radiologists. Our findings suggest such AI-generated efficiencies in image analysis could create $3 billion in annual savings by giving radiologists more time to focus on reviews that require greater interpretation or judgment. Another area is AI-assisted robotic surgery. In **orthopedic** surgery, a form of AI-assisted robotics can analyze data from pre-op medical records to physically guiding the surgeon's instrument in real time during a procedure. It can also use data from actual surgical experiences to inform new surgical techniques. A study of 379 orthopedic patients across nine surgical sites found that an AI-assisted robotic technique created by Mazor Robotics resulted in a fivefold reduction in surgical complications compared with when surgeons operated alone. When it is applied properly to orthopedic surgery, our analyses find AI-assisted robotic surgery could also generate a 21% reduction in patients' length of stay in the hospital following surgery, as a result of fewer complications and errors, and create $40 billion in annual savings.

5 AI techniques are also being applied to the costly problem of **dosage** errors—where our findings suggest AI could generate $16 billion in savings. In 2016, a groundbreaking trial in California found that a mathematical formula developed with the help of AI had determined the correct dose of **immunosuppressant** drugs to administer to organ patients. Determining the dose has traditionally depended on a combination of guidelines and educated guesswork—and dosing errors make up 37% of all preventable medical errors. While this type of AI technique is **nascent**, the example is powerful considering that the correct dose is critical to making sure a graft is not rejected after an organ transplant.

6 Using AI to aid clinical judgment or diagnosis still remains in its infancy, but some results are emerging to illustrate the possibility. In 2017, a group at Stanford University tested an AI algorithm against 21 **dermatologists** on its ability to identify skin cancers. The clinical findings, as reported by *Nature* last year, "achieve performance **on par with** all tested experts…demonstrating an artificial intelligence capable of classifying skin cancers with a level of competence comparable to dermatologists". Our findings suggest AI could **yield** $5 billion in annual savings by doing a preliminary diagnosis before a patient enters the emergency department.

7 We're also starting to see the potential of AI-powered virtual nurse assistants in helping patients. For example, Sensely's Molly is an AI-powered nurse **avatar** being used by UCSF and the U.K.'s NHS to interact with patients, ask them questions about their health, assess their symptoms, and direct them to the most effective care setting. Our findings estimate AI-powered nurse assistants could save $20 billion annually by saving 20% of the time nurses spend on patient maintenance tasks.

8 AI also holds promise for helping the healthcare industry manage costly back-office problems

and inefficiencies. Activities that have nothing to do with patient care consume over half (51%) of a nurse's workload and nearly a fifth (16%) of a physician's activities. AI-based technologies, such as voice-to-text transcription, can improve administrative workflows and eliminate time-consuming non-patient-care activities, such as writing chart notes, filling prescriptions, and ordering tests. We estimate that these applications could save the industry $18 billion annually.

9 For example, while Beth Israel Deaconess Medical Center garnered attention for an AI-enabled cancer screen, its first foray into AI was more prosaic: using it to reduce hospital readmission rates and identify possible no-shows. Using machine learning, technologists there developed an application to predict which patients are likely to be no shows or lapse on treatment so they can intervene ahead of time.

10 Errors and fraud are a similarly expensive problem for healthcare organizations and also for insurers. Fraud detection has traditionally relied on a combination of computerized (rules-based) and manual reviews of medical claims. It's a time-consuming process that **hinges on** being able to quickly spot anomalies after the incident occurs in order to intervene. Health insurers are experimenting with AI-supported data mining, coupled with AI-based neural networks (which mimic the processes of the human brain, but much more quickly) to search Medicare claims for patterns associated with medical **reimbursement** fraud. We estimate that AI could create $17 billion in annual savings by improving the speed and accuracy of fraud detection in Medicare claims.

11 Beyond fraudulent activity, the **litany** of data **breaches**, such as WannaCry or Petya, over the past few years has made cybersecurity a major concern for healthcare organizations. Healthcare breaches are estimated to cost organizations $380 per patient record. Using AI to monitor and detect abnormal interactions with proprietary data could create $2 billion in annual savings by reducing health record breaches.

12 As AI technologies become more prevalent, healthcare organizations will have to invest in those that deliver the most value. The use of AI for clinical judgment still remains in its infancy and will need time to fully **take root** in a meaningful way. But the AI applications that can deliver the most value today (AI-assisted surgery, virtual nurse, administrative workflow, etc.) should be prioritized and invested in, so healthcare providers and insurers are free to focus on better care.

Notes

Rock Health a seed fund investing in digital health start-ups to support entrepreneurs working at the intersection of healthcare and technology.

NYU Langone Health New York University Langone Health, based in New York City, one of the nation's premier academic medical centers devoted to patient care, education, and research.

Mazor Robotics an Israeli medical device company and manufacturer of a robotic guidance system for spine surgery, a pioneer and leader in the field of surgical guidance systems.

Nature the world's leading multidisciplinary science journal first published in 1869.

UCSF University of California, San Francisco, located in San Francisco, California, a unique branch of the University of California system that focuses exclusively on medicine and life sciences.

NHS the British National Health Service system, which always shoulders the heavy responsibility of safeguarding public healthcare in the United Kingdom, complies with the principle of selectivity for the relief of the poor, and advocates the principle of universality.

Words and Expressions

crop up to appear suddenly or unexpectedly 突然出现

streamline /ˈstriːmlaɪn/ *v.* to make a system, an organization, etc. work better, especially in a way that saves money 使（系统、机构等）效率更高;（尤指）使增产节约

clinician /klɪˈnɪʃn/ *n.* a doctor, psychologist, etc. who has direct contact with patients 临床医师

aptitude /ˈæptɪtjuːd/ *n.* natural ability or skill at doing something 天资，天赋

anomaly /əˈnɒməli/ *n.* a thing, situation, etc. that is different from what is normal or expected 异常事物，反常现象

radiologist /ˌreɪdiˈɒlədʒɪst/ *n.* a doctor who is trained in radiology 放射科医生；X 光科医生

nodule /ˈnɒdjuːl/ *n.* a small round lump or swelling, especially on a plant （尤指植物上的）结节，小瘤

orthopedic /ˌɔːθəˈpiːdɪk/ *adj.* relating to the medical treatment of problems that affect a people's bones or muscles 矫形外科的

dosage /ˈdəʊsɪdʒ/ *n.* the amount of a medicine or drug that you should take at one time, especially regularly （药的）剂量，服用量

immunosuppressant

/ˌɪmjʊnəʊsəˈpresənt/ *n.* a drug that reduces the body's natural immunity by suppressing the natural functioning of the immune system ［药］免疫抑制剂（素）

nascent /ˈnæsnt/ *adj.* beginning to exist; not yet fully developed 新生的，萌芽的；未成熟的

dermatologist /ˌdɜːməˈtɒlədʒɪst/ *n.* a doctor who studies and treats skin diseases 皮肤病医生，皮肤病专家

on par with as good, bad, important, etc. as somebody/something else 与……同等水平

yield /jiːld/ *v.* to produce or provide something, for example, a profit, result, or crop 产生（收益、效益等）；出产（作物）；提供

avatar /ˈævətɑː(r)/ *n.* (in Hinduism and Buddhism) a god appearing in a physical form 化身（印度教和佛教中化作人形或兽形的神）

hinge on to depend on something completely 取决于；有赖于

reimbursement /ˌriːɪmˈbɜːsmənt/ *n.* compensation paid for damages or losses or money already spent 报销；偿还；赔偿

litany /ˈlɪtəni/ *n.* a long boring account of a series of events, reasons, etc. （对一系列事件、原因等）枯燥冗长的陈述

breach /briːtʃ/ *n.* an action that breaks an agreement to behave in a particular way 破坏；辜负

take root to become settled and stable in one's residence or lifestyle 生根，扎根

Useful Terms

venture capital firm 风险投资公司

lung nodule 肺结节

surgical complication 手术并发症

back office 后勤部门

Exercises

Comprehension Check

I. Answer the following questions according to the text.

1. How are these top ten AI applications in healthcare selected?
2. In what areas does AI currently create the most value?
3. How can AI technologies help radiologists?
4. What does AI-assisted robotics do in orthopedic surgery?
5. In what ways can AI-powered virtual nurse assistants help patients?

II. Read the following statements carefully and decide whether they are true (T) or false (F) without turning back to check the text.

1. ______________ The value of AI applications can create up to $150 billion by 2026 in 206 deals in healthcare.
2. ______________ AI creates the most value in making clinical decisions and improving clinical outcomes.
3. ______________ In orthopedic surgery, AI-assisted robotics shortens patients' stay in hospital.
4. ______________ AI techniques help reduce the patients' dosing errors.
5. ______________ Using AI to aid clinical judgments or diagnosis has reached its maturity.

III. Choose the best answer to each of the following questions according to the text.

1. People are excited about the value of AI applications in healthcare for the following reasons EXCEPT that ______________.

   A. AI contributes a lot in different areas of healthcare

   B. AI helps people streamline administrative and clinical healthcare processes

   C. AI-assisted clinical applications have been ripe enough

   D. AI is expected to generate more revenue in many fields

2. According to Figure 7.1, AI has been applied in ______________ to create the most value.

   A. robot-assisted surgery　　B. virtual nurse assistants

C. fraud detection　　　　D. cybersecurity

3. The applications of AI in healthcare benefit people in the following aspects EXCEPT that ______________.

   A. they could create revenue in annual savings for U.S. healthcare

   B. health AI raised a great fund in 206 deals between 2011 and 2017

   C. AI-assisted robotics can analyze data to guide surgeons' instrument in real time

   D. AI detects and matches lung nodules faster than a group of radiologists

4. Which of the following statements is NOT true?

   A. AI's aptitude in identifying skin cancers can be comparable to that of demonologists.

   B. AI-assisted clinical diagnosis is commonly seen globally.

   C. Molly is an AI-powered nurse avatar to interact with patients and provide them with the most effective care setting.

   D. AI contributes a lot to improving the speed and accuracy of fraud detection in medical claims.

5. Which area of healthcare does NOT deserve to be prioritized and invested most?

   A. AI-assisted surgery.　　　　B. Virtual nurse.

   C. Administrative workflow.　　　　D. Clinical judgment.

Vocabulary Building

IV. Fill in the following blanks with the words and phrases given in the box. Change the form if necessary.

aptitude	dosage	eliminate	preliminary	yield
clinical	hinge on	anomaly	on par with	physician

1. Nonetheless, the county's population grew roughly ______________ the rest of the nation.

2. The space shuttle had stopped transmitting data, a very serious ______________ for the mission.

3. In this article, I describe how to ______________ the redundant data in each row, then merge the cells together, and display the data on the cells.

4. Yet what he did or did not want to do often seemed to ______________ what I might

learn from doing it myself.

5. Unless you're just starting out in life, you have some skills or talent to show some kind of ______________.

6. What can we do next to ensure its effectiveness in the following ______________ trials?

7. However, you may reduce the ______________ with the advice of doctor.

8. Each of these oilfields could ______________ billions of barrels of oil every year.

9. As yet, we have only ______________ thoughts on the first question and none to report on the second.

10. ______________ skill in these areas increases the likelihood that patients will have more realistic expectations regarding treatment outcomes.

V. Match the words in the left column with the explanations in the right column.

1.	efficiency	A.	innovative, pioneering, original
2.	surgical	B.	to make a judgment about the nature or quality of somebody/something
3.	groundbreaking	C.	the quality of doing something well with no waste of time or money
4.	investigate	D.	relating to or used for medical operations
5.	assess	E.	to carefully examine the facts of a situation, an event, a crime, etc.

Word Formation

VI. Fill in the following blanks with the words in capitals. Change the form if necessary. An example has been given.

e.g. *I am a little nervous about leaving the kids at home all alone.* **NERVE**

1. It was ______________ of her to jump into the water to save the drowning boy. **COURAGE**

2. It would be ______________ to live in a peaceful world. **GLORY**

3. Tom, like many ______________ young lawyers, became interested in politics. **AMBITION**

4. The book is a ______________ account of a young man's travels in Asia. **HUMOR**

5. Lack of money puts us in a ______________ and unfavorable position. **DISADVANTAGE**

Translation

VII. Translate the following sentences into English with the words and phrases in brackets.

1. 人工智能在医疗领域的应用前景看似广阔，涵盖了健康、诊断和操作技术等内容，但实际上它很狭窄，因为这些应用仅用于执行单项任务。(in that; perform)

____

____

2. 创业公司 Sensely 开发了一个名为 Molly 的数字护士，以帮助医生监测患者在就诊期间的病情，并跟踪治疗情况。(monitor; follow up)

____

____

3. 人工智能系统可以用来分析数据，这些数据来自患者档案、外部研究及临床专业知识的注释和报告，以帮助人们选择正确的、单独定制的治疗方法。(external research; clinical expertise)

____

____

Text B

The Eye of Aritificial Intelligence in China

1 "Mirror mirror on the wall, who is the most beautiful in the world?" When old tales tell a seemingly unbelievable story of a mysterious mirror, modern technology secretly turns the magic into reality.

2 In August, news **went viral** that the police had caught criminal suspects who attended Chinese singer Jacky Cheung's concerts. How did the police pick out the suspects among thousands of screaming fans? Well, the mysterious helping hand is AI facial recognition technology. The low-key yet powerful tool is no longer far-fetched, but entering the world of China's top investors and start-up companies and even me and you.

Technical Principles: How Does AI Facial Recognition Work?

3 Facial recognition is a technology capable of identifying or verifying a person from a digital image or a frame from a video source. There are multiple methods in which facial recognition systems work, but in general, they work by comparing selected facial features from given images with faces within a database.

4 Tang Wenbin, CTO of Face++, an AI computer vision technology company, vividly explained the **workflow** of AI. According to him, facial recognition is the identification of people's appearance, including knowing the gender, age, and identity. You may think it is like a brain or a neural network. You use data to train it; then, it learns the pattern. The challenge is using less data to build a better model and to compute a better algorithm.

Applications—Security, Mobile Phones, Finance, and Beyond

5 Facial recognition technology has been traditionally associated with the security sector, but today there is an active expansion into other industries including retail, marketing, and finance. By 2022, the global facial recognition technology market **is projected to** generate an estimated $9.6 billion in revenue with a compound annual growth rate of 21.3%, according to MarketWatch. In the security area, AI cameras are installed in cities to work day and night to help recognize wanted criminals. When it spots a face in a mass database, the nearest police office is alerted, and officers can rush to the scene and pursue the criminals. The whole process can take as little as 20 minutes.

6 Meanwhile, China has issued the Cybersecurity Law of the People's Republic of China to protect the online security and privacy of citizens. The law states that network operators shall establish and complete user information protection systems and strictly preserve the secrecy of user information they collect.

7 China has also developed a fast and convenient 3D facial unlock function with the new 3D sensor and leading facial recognition technique to apply the technology to mobile phones.

8 AI can also bring convenience to the finance sector. Facial recognition is the most efficient security means in opening a bank account, transferring money, and shopping and paying online because it is the most convenient tool.

Business: From Rapid Acquisitions to Fierce Competitions

9 Chinese start-ups have seen AI facial recognition's potential to simplify and speed up tasks in multiple industries. And the market is now booming, crowded, and competitive.

10 The Chinese market is more competitive than that of the U.S. If your product is only 1% more accurate than the competitor's, you could hold the most market share, according to Tian Feng, director of Alibaba Cloud Research Center.

11 Chinese start-up companies went through three stages. According to Yin Qi, CEO of Face++, from 2013 until now, most of the companies have been focusing on developing core technology. In the second stage, especially from 2017, they have been focusing on the project of AI and industry. Yin Qi has deemed that the combination of AI and the Internet is a big opportunity. However, in 2018, when AI is applied to different industries, he thinks the industry enterprises have occupied is like a bargaining chip. Whether you have one or two of those chips that can participate in the next stage, development is the key.

12 "When you delve into the industry, you have to divide and **reallocate** the profit with other enterprises. But I think the potential of this industry is essential and the chain of supply and demand will certainly change," Yin Qi said.

Big Investors Are Placing Bets on China's Facial Recognition Start-ups

13 China's facial recognition start-ups are attracting huge investments. In July, Chinese facial recognition companies, according to a pair of reports, were raising as much as $1.6 billion. Those investments would build on billions of dollars that investors have already put into the companies.

14 But not all AI companies are **under the** investment **spotlight**. Sheng Xitai, founder of Hongtai Aplus and chairman of Hongtai Capital Holdings, lists the conditions that he finds attractive. He thinks commercialization is one of the crucial points. And there are a couple of important points. First, the company has to have a solid technology background. Second, it needs to find a field that is narrow and deep enough that no other big companies have paid attention to. Third, the company should have to look for a suitable case to apply its technology. Fourth, it has to develop verifiable customers, not the **pseudo-conceptual** ones. Finally, the company needs to develop loyal buyers. That will lead to sustainable development.

15 China is embracing the AI revolution and has gained a breakthrough resulting in facial recognition. However, the focus of AI research in China is a bit different than that of U.S. Yin Qi thinks the research is at the same level as the U.S., but the focus is a bit different. China's application of innovation is more developed than that of other countries, but from the perspective of technology, we need more research in essential fields, such as computer vision and deep learning.

16 But in general, the blueprint and prospects for the application of AI are seen to be bright for many industries. If applying AI to different industries is a **marathon**, the companies probably have only run one or two kilometers. Tian Feng also thinks that in the future, everybody will be able to use AI as it will soon become a basic tool for work and life. For an investor like Sheng Xitai, the industry of AI has just begun. The number of fields that it has been applied to is incomparable to the Internet and has unlimited possibilities.

Notes

Jacky Cheung born in Hong Kong on July 10, 1961, a singer and actor in China.

MarketWatch a financial information website owned by Dow Jones, which focuses on news and analysis on stocks, business, and politics. The target of the service is more for retail investors.

Cybersecurity Law of People's Republic of China
a law enacted to safeguard network security, maintain cyberspace sovereignty, national security, and social public interests, protect the legitimate rights and interests of citizens, legal persons, and other organizations, and promote the healthy development of economic and social information.

Alibaba know as Alibaba Network Technology Co., Ltd., which is a company that mainly provides online trading platform for e-commerce. It was founded in 1999 in Hangzhou, Zhejiang Province by 18 people headed by Jack Ma, an English teacher.

Hongtai Aplus founded in 2014 by veteran banker Sheng Xitai and New Oriental founder and CEO Michael Yu. It invests in the consumer, healthcare, education, and mobile Internet sectors in China.

Words and Expressions

go viral to quickly and widely circulate on the Internet, as of a video, picture, or post 疯狂传播；像病毒般扩散；走红

workflow /ˈwɜːkfləʊ/ *n.* the way that a particular project is organized by a company, including which part of a project somebody is going to do, and when he/she is supposed to do it 工作流程

be projected to to be planned for the future 预计

acquisition /ˌækwɪˈzɪʃn/ *n.* a company, piece of land, etc. bought by somebody, especially another company; the act of buying it 收购，购置；购置物；收购的公司；购置的产业

reallocate /ˌriːˈæləkeɪt/ *v.* to change the way money or materials are shared between different people, groups, projects, etc. 重新分配，再分配

under the spotlight under the focus of public attention 在聚光灯下

pseudo-conceptual /ˌsuːdəʊ kənˈseptʃuəl/
adj. a false or spurious concept 伪概念的

marathon /ˈmærəθən/ *n.* a long running race of about 42 kilometers or 26 miles 马拉松赛跑（距离约 42 公里，合 26 英里）

Useful Terms

facial recognition 人脸识别

mass database 海量数据库

rapid acquisition 快速收购

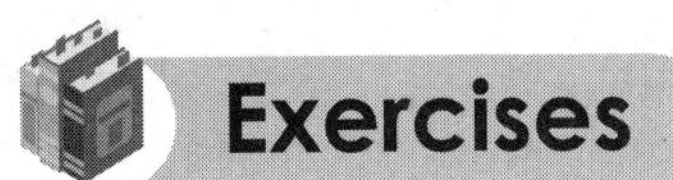

Exercises

Comprehension Check

I. Identify the paragraph from which the information is derived.

1. ______________ Facial recognition has been expanded into other industries.
2. ______________ Facial recognition is the most efficient security means in the finance sector.
3. ______________ AI facial recognition technology helped the police catch criminal suspects at a singer's concerts.
4. ______________ Only those companies that are under five favorable conditions can attract investors.
5. ______________ Facial recognition works by comparing the selected facial features of given images with faces in a database.
6. ______________ In the near future, the facial recognition market will generate great revenues.
7. ______________ AI applications have a bright future ahead.
8. ______________ China has been focusing on the development of core technology for years.
9. ______________ Facial recognition works like a brain or a neural network.
10. ______________ AI cameras work all day helping recognize wanted criminals.

II. Answer the following questions according to the text.

1. What does the word "eye" refer to in the title "The Eye of Artificlal Intelligence in China"?
2. How many applications of AI in facial recognition are mentioned ?And what are they?

3. What kind of AI companies can attract more investment according to Shen Xitai?

Vocabulary Building

III. Fill in the following blanks with the words and phrase given in the box. Change the form if necessary.

suspect	low-key	expansion	project	compound
alert	privacy	verifiable	sustainable	under the spotlight

1. I ______________ that there was something wrong with the engine.
2. All human power is a(n) ______________ of time and patience.
3. But when we ______________ ever in the future, we forget everything we have.
4. We need to have a more stable and ______________ business model because, candidly, we only want to do this once.
5. We only know it exists because of its effect on the ______________of the universe.
6. Instead, we should start worrying about who will be the next victim of these ______________ violations. It could be anyone.
7. But you have to be very ______________ when you are most vulnerable, depressed, or isolated.
8. This ______________ launch is designed to support the idea that the new features are merely "experimental".
9. This week we put these issues ______________ in a series of articles.
10. Is all of the above information documented and ______________?

Aritificial Intelligence Brings High Returns on Drug Discovery and Development

Machine learning has broad applications in healthcare where the availability of rich, well defined data sets, the need for monitoring over time, and the wide variability of outcomes offer the potential for disproportionately high returns on the technology implemented in areas like drug discovery, test analysis, treatment optimization, and patient monitoring. With the integration of machine learning and AI, the opportunity exists to significantly derisk the drug discovery and development process, removing $26 billion per year in costs, while also driving efficiencies in healthcare information worth more than $28 billion per year globally.

What Are the Opportunities?

Drug Discovery and Development

The potential efficiency gains from incorporating machine learning processes throughout development could not only accelerate the time horizon but also improve returns on R&D spend by increasing the probability of success (POS) for medicines reaching late-stage trials. According to David Grainger, partner at Medicxi Ventures, avoiding the false discovery rate, a mostly statistical driven phenomenon according to him, could derisk late-stage trials by half. Further, the current method of virtual screening in early-stage drug discovery known as high throughput screening (HTS) is highly **vulnerable** to this type of statistical error. Halving the risk of expensive Phase Ⅲ trials could generate billions in savings and meaningfully impact returns on the more than $90 billion in R&D spend across the largest pharmaceutical companies, freeing up resources to focus on finding higher potential opportunities.

v. 易受攻击的

While substantial costs associated with late-stage trials often focus on clinical trial design elements, we believe meaningful efficiency gains can also be realized throughout later stages with AI/ML implementation to optimize decisions around selection **criteria**, size, and length of study.

n. 标准

Doctor/Hospital Efficiency

Driven partly by regulation and fragmentation, the healthcare system in the U.S. has historically been slow to adopt new technologies. Apart from the systematic challenges, the time between new discoveries and when doctors and clinics put new medicines or treatments to use is often long and **inconsistent.**

adj. 不一致的

Recent mandates from the U.S. government as part of the American Recovery and Reinvestment Act have driven growth in spaces like electronic health records, a global market expected to reach around $30 billion by 2023, according to Transparency Market Research(see Figure 7.2). The **aggregation** of data, improving technology to capture it, and **secular** decline of stand-alone hospitals has created an opportunity to leverage data at a scale not attainable historically. This in turn is enabling machine learning algorithms and AI capabilities to demonstrate early traction improving the speed, cost, and accuracy in various areas of healthcare.

n. 聚合，集合
adj. 长期性的

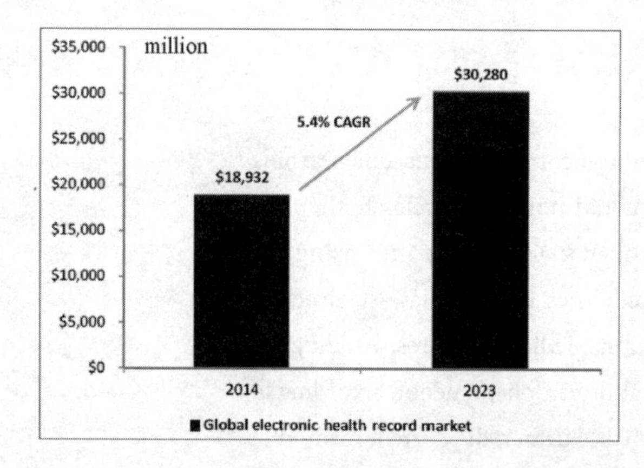

Figure 7.2 Electronic health records market expanding globally

Google's DeepMind division headquartered in London is collaborating with the U.K.'s National Health Service (NHS) to build an app aimed at monitoring patients with kidney disease as well as a platform formerly known as "Patient Rescue" that seeks to support diagnostic decisions. A key input for any AI/ML system is immense amounts of data, so DeepMind and the NHS entered a data-sharing agreement providing

DeepMind with a continuous stream of new data and historical records it is leveraging to train its algorithms. This real-time analysis of clinical data is only possible with vast amounts of data, though the insights provided by DeepMind's effectively unlimited access to patient data could deliver learnings well beyond the scope of kidney disease.

What Are the Pain Points?

Drug Discovery and Development

A significant pain point within healthcare is the time and cost of drug discovery and development. It takes approximately 97 months on average for new therapies to progress from discovery to Food and Drug Administration (FDA) approval, according to the Tufts Center for the Study of Drug Development (TCSDD). While a focus on specialty can aid the time horizon, costs have continued to increase steadily as well. Deloitte found that across a **cohort** of 12 major pharmaceutical companies, the cost to develop an approved asset has increased 33% since 2010 to roughly $1.6 billion per year.

n. 一群

R&D Returns

R&D productivity in biopharma remains a debated topic. While the cost of developing successful drugs has increased, the revenue environment has also been unsupportive of returns due to reimbursement headwinds, lower patient volumes, and competition. While we expect returns to improve from 2010–2020 vs 2000–2010, the change is marginal (see Figure 7.3). Further, one of the most significant headwinds to returns remains failed assets, particularly those reaching later stages, which we estimate account for more than $40 billion in annual costs.

Total return ratio (cumulative revenue divided by GAAP R&D spend)

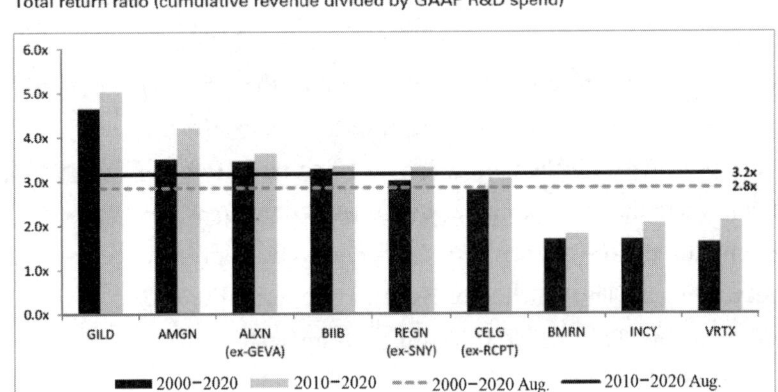

Figure 7.3 20-year and 10-year total return ratios
(Source: Company data, Goldman Sachs Global Investment Research)

Doctor/Hospital Efficiency

A challenge unique to healthcare remains the significant **lag** between when new drugs and treatments are approved versus when doctors begin implementing with patients. As a result, many machine learning and AI experts working in healthcare continue to encourage major providers to integrate modern machine learning tools into their workflows which can fully utilize the vast stores of medical data being collected and published today.

n. 时滞，延迟

Opportunities exist for machine learning and AI to decrease the time between discovery and application and also optimize treatments. For example, a 2009 study from the Radiological Society of North America on **hepatobiliary** radiology found that 23% of second opinions had a change in diagnosis, a problem machine learning companies focused on medical imaging have an opportunity to solve. Further, companies like Deep Genomics that use machine learning to identify diseases at the genome level are positioning providers to deliver more targeted and effective treatments.

adj. 肝胆的

What Is the Current Way of Doing Business?

The current drug discovery and development business is an extensive process of research, testing, and approval that can last more than ten years. Time to market analysis from the TCSDD reports that it takes 96.8 months on average for a drug to advance from Phase Ⅰ to FDA approval. Discovery of new treatments is a unique challenge not only because of the length of time required but also because of the low POS throughout the various stages of development.

Drug discovery initially begins with identifying a target. Once a target has been identified, HTS is often used for "hit finding". HTS is an **automated**, expensive process carried out by robots that tries to identify these "hits" by conducting millions of tests to see which compounds show potential with the target. The hits then **transition** to lead generation where they are optimized to find lead compounds, which are then optimized more extensively before progressing to pre-clinical drug development. This entire process can last one to three years before a drug reaches Phase Ⅰ, at which time it is understood to have only a 20% probability of success.

adj. 自动化的

v. 过渡；转变

- Phase Ⅰ: Emphasis on safety; healthy volunteers (20% POS).
- Phase Ⅱ: Focuses on effectiveness; volunteers with certain diseases or conditions (40% POS).
- Phase Ⅲ: Further information gathered on safety and effectiveness across different populations, dosages, and combinations; ranges from several hundred to thousands of volunteers (60% POS).

How Does AI/ML Help?

The advantages and use cases of machine learning and AI within the healthcare industry are wide-ranging. Not only are decisions driven by data, rather than by human understanding or **intuition**, but the decisions and/or predictions are able to consider a combination of factors beyond human capacity. Deep learning in particular shows unique potential as it can exploit knowledge learned across different tasks in order to improve performance on others tasks.

n. 直觉；判断力

AI/ML can reduce failed discovery and increase POS. Significant capital is invested with substantial opportunity cost in exploring treatments that are understood to have roughly 20% POS if they reach Phase Ⅰ trials. As a result, AI/ML has been applied, almost entirely within academics to date, in an effort to develop efficient and accurate virtual screening methodologies to replace costly and time-consuming HTS processes.

AI/ML improve doctor/hospital efficiency. Early successes in applying machine learning have been seen with improving diagnoses (Enlitic, DeepMind Health), analyzing **radiology** results (Zebra Medical Vision, Bay Labs), genomic medicine (Deep Genomics), and even the use of AI in treating depression, anxiety, and PTSD (Ginger.io). As health data becomes more accessible as a result of both the digitization of health records and aggregation of data, significant opportunity exists for AI/ML to not only remove costs associated with procedural tasks but also improve care via algorithms that let historically disparate data sets communicate. Ultimately, the capabilities of AI/ML to consider factors and combinations beyond human capacity will allow providers to diagnose and treat with greater efficiency.

n. 放射学

Comprehension Check

I. Answer the following questions according to the text.

1. Why are AI/ML applications so wide-ranging in healthcare?
2. How can the potential efficiency be gained in drug discovery and development?
3. Why do the new technologies, medicines, or treatments have to go a long way before they are used in the U.S.?
4. What are the pain points of AI/ML applications in healthcare?
5. How does AI/ML function in the process of drug discovery and development?

II. Read the following statements carefully and decide whether they are true (T) or false (F) without turning back to check the text.

1. ______________ AI and ML work together to reduce the costs of drug discovery and development process greatly as well as driving efficiencies in healthcare information globally.
2. ______________ According to David Grainger, avoiding the false discovery rate, a mostly statistical driven phenomenon according to him, could derisk late-stage trials by 50%.
3. ______________ The current drug discovery and development business is an extensive process of research, testing, and approval that can last less than ten years.
4. ______________ The discovery of new treatments needs long time because of the low POS throughout the various stages of development.
5. ______________ Although AI/ML technologies have been applied to healthcare significantly, it is human being's understanding and intuition that decide on the clinical diagnosis and medical treatment.

III. Discuss the following questions based on your understanding of the applications of artificial intelligence.

1. What is the prospect of AI applications in China's healthcare industry?
2. What problems does AI bring to medical ethnics?

3. Is it possible that the discovery and development of new medicines and new treatments will totally rely on the AI/ML technologies?

Writing Skills

非谓语动词

在英语写作中，非谓语动词（non-finite verbs）的巧妙使用会使写作质量显著提高，但这同时也是一个写作难点。只有花时间下功夫掌握它，才能在英语写作中游刃有余。

非谓语动词是指在句子中不作谓语的动词，包括不定式、动名词和分词三种形式。除了不能作谓语外，非谓语动词可以充当句子中的很多成分，如状语、定语、表语等。

使用非谓语动词时需要注意以下三点：第一，动名词作句子主语或表语；第二，动词不定式主要作目的状语；第三，分词（现在分词和过去分词）可以作定语、表语、宾语补足语，也可以作状语。

接下来我们着重介绍分词的使用，包括如何把并列句转化为其他句式，或者如何从熟悉的从句入手进行句式的转化。例如：

- **并列句的转化：** She stood by the window and looked at the sunset. → Standing by the window, she looked at the sunset. / She stood by the window, looking at the sunset.
- **定语从句的转化：** The boy who is sitting under the tree is my cousin. → The boy sitting under the tree is my cousin.
- **状语从句的转化：** Don't laugh when you are having dinner. → Don't laugh when having dinner.

需要注意的是，当分词作状语时，其逻辑主语需要与句子的主语保持一致。分词作状语的类型主要有以下五种：

- **时间状语：** Studying in Beijing University, he met his first girlfriend.
- **原因状语：** Being spring, the flowers are in full bloom.
- **结果状语：** The old scientist died all of a sudden, leaving the project unfinished.
- **条件状语：** It will take you an hour to get to the airport, allowing for traffic delays.
- **让步状语：** Granting this to be true, we cannot explain it.

Rewrite the following sentences using participles as adverbials.

1. In the security area, AI cameras are installed in cities to work day and night to help recognize wanted criminals.

____

____

2. Facial recognition is the most efficient security means in opening a bank account, transferring money, and shopping and paying online because it is the most convenient tool.

____

____

3. If applying AI to different industries is a marathon, the companies probably have only run one or two kilometers.

____

____

4. On many targets it achieves nearly perfect prediction quality, which qualifies it for usage as a virtual screening device.

____

____

5. Significant capital is invested with substantial opportunity cost in exploring treatments that are understood to have roughly 20% probability of success (POS) if they reach Phase Ⅰ trials.

____

____

Translating Skills

拆句法与合并法

拆句法（splitting）和合并法（combination）是两种相互对应的翻译方法。前者是把一个长而复杂的句子拆译成若干个较短、较简单的句子，通常用于英译汉；后者是把若干个短句合并成一个长句，一般用于汉译英。汉语强调意合，结构较松散，因此简单句较多；英语强调形合，结构较严密，因此长句较多。所以，汉译英时需要利用连词、分词、介词、不定式、定语从句、独立结构等把汉语短句合并成长句；英译汉时通常需要在原句的关系代词、关系副词、主谓连接处、并列或转折连接处、后续成分与主体的连接处，以及意群结束处将长句切断，译成汉语分句。这样可以在基本保留英语语序的基础上，顺译全句，并符合现代汉语长短句相替、单复句相间的句法修辞原则。

例 1：We identified these specific AI applications based on how likely adoption was and what potential exists for annual savings.
根据采用的可能性以及每年节省成本的潜力，我们确定了这些特定的人工智能应用程序。（在分词处拆译）

例 2：AI has demonstrated its aptitude for improving the efficiency of image analysis by quickly and accurately flagging specific anomalies for a radiologist's review.
人工智能通过快速、准确地标记特定异常以供放射科医师评估，证明了它在提高图像分析效率方面的能力。（在介词处拆译）

例 3：一项针对 379 名骨科患者九个手术部位的研究发现，与外科医生单独手术相比，Mazor Robotics 创建的人工智能辅助机器人技术使手术并发症减少了 80%。
A study of 379 orthopedic patients across nine surgical sites found that an AI-assisted robotic technique created by Mazor Robotics resulted in a fivefold reduction in surgical complications compared with when surgeons operated alone.（用宾语从句、分词合译）

例 4：2016 年，加利福尼亚州的一项突破性试验发现，在人工智能帮助下开发的一个数学公式，可以确定器官患者所需免疫抑制药物的正确剂量。
In 2016, a groundbreaking trial in California found that a mathematical formula developed with the help of AI had determined the correct dose of immunosuppressant drugs to administer to organ patients.（用介词、宾语从句、分词合译）

Translation at Sentence Level

I. Translate the following sentences into Chinese.

1. They can send alerts to the users to get more exercise and share this information with doctors (and AI systems) for additional data points on the needs and habits of patients.

____

____

2. Precision insurance uses genome-based algorithmic insurance focused on susceptibility to areas such as physical diseases, mental health risks, and the future cost of delayed-onset diseases.

____

____

3. The next step is to launch a large-scale user study where the technology will be deployed across cancer registries to identify the most effective ways of integration in the registries' workflows.

____

____

4. IEI strives to provide the powerful digital hospital solutions, such as nursing station, mobile nursing terminals, and smart bedside infotainment to improve patient-doctor relationship by assisting caregivers to provide necessary healthcare education for inpatients before and after treatments.

____

____

5. Regulation and legislation will be vitally important to reassure patients that AI is part of a wider network which is human and ultimately responsible and accountable for what happens to them.

____

____

II. Translate the following sentences into English.

1. 2016年，波士顿儿童医院为亚马逊Alexa开发了一款应用程序。这款应用程序为患病儿童的父母提供基本的健康建议。

____

2. 在最近由埃博拉病毒引发的疫情中，一个由人工智能驱动的程序可以扫描现有的药物，这些药物经过了重新设计来对抗这种疾病。

3. 众多的深度学习模型可以检测和分类影像学，其性能与放射科医师相当。

4. 我们的结果表明，相比于患者群体、疾病特征和成像系统中具有不同数据分布的数据集，有限的数据集中训练的深度学习模型表现不好。

5. 通过与美国国家癌症研究所进行合作，我的团队提出了先进的人工智能解决方案，即自动采集耗时的数据和提供近乎实时的癌症报告，使国家癌症监测计划实现现代化。

Translation at Paragraph Level

English to Chinese Translation

①Dr. Topol in his book *Deep Medicine* proposes that artificial intelligence has the potential to free physicians from the logistical tasks that interfere with their ability to connect to patients. ②Contrary to concerns that tech is actually distancing patients from healthcare—commentators have expressed concerns that new-fangled interventions inadvertently isolate older patients, that people do not want to accept diagnoses from AI, and that prejudices within an algorithm can prevent care from reaching the right people—Topol and others like him believe AI can actually help make healthcare more human. ③Anyone who has kept his finger on the pulse of medical trends in recent years will be well aware that AI has the power to revolutionize the way healthcare is delivered across the board. ④The area where it has the potential to be the most transformative is also, arguably, the least exciting—administration.

本段一共四句话。第一句涉及从句的翻译，需要分译。根据第二个从句对它的进一步阐述，logistics tasks 可以译为"后勤任务"，从而更加符合汉语的表达习惯。所以整句可以译为："Topol 博士在他的《深层医学》一书中提出，人工智能可以把医生从后勤任务中解放出来，这些后勤任务会影响他们与患者的联系能力。"

第二句比较长，需要注意的是 concerns 后面有三个由 that 引导的同位语从句，用来解释说明这些"担忧"有哪些。此句可以译为："与技术实际上使患者疏离医疗保健的担忧相反——评论员担忧的是，新型干预措施无意中将老年患者隔离开来，患者不希望接受人工智能的诊断，以及算法中的偏见可能会阻止将护理提供给所需的患者——Topol 和其他同伴相信人工智能实际上可以使医疗保健变得更加人性化。"

第三句中的 be aware that 后接的宾语从句也需要分译。除此之外，习语 keep one's finger on the pulse 的意思是"紧跟动态"。所以整句可以译为："任何一个对近年来医疗趋势保持关注的人都会清楚地意识到，人工智能有能力彻底改变医疗保健的方式。"

第四句最短，但也最难翻译。arguably 的翻译是难点，这里根据上下文语境可以理解为"按理说"。这句话的意思是："可以说，管理既是最具变革潜力的领域，也是最令人不感兴趣的领域。"

III. Translate the following paragraph into Chinese.

Domestic Internet giant Baidu recently announced that its public welfare Baidu Artificial Intelligence People Searching program successfully reunited more than 10,000 lost people with their families over the past three years, which is a good example of how technology benefits people. Baidu launched this project using its AI-based facial recognition technology. In cooperation with the Ministry of Civil Affairs and non-governmental organizations, Baidu's system matches the photographs provided by seekers with those in the population database to achieve highly efficient search results. After going through 200 million face-training sample photos, Baidu's AI-based facial recognition technology has achieved a recognition accuracy rate of as high as 99.7%. It also supports across-age image comparisons, thus helping parents find their lost children after several years on the basis of their childhood photos.

____

____

____

____

____

____

Chinese to English Translation

①从病人的可穿戴技术到辅助医生进行手术的技术，人工智能已经是医学领域的一个关键角色。②随着人工智能市场的总体增长量预计在未来六年里会以 63% 的复合年增长率达到 166 亿美元，这一领域正在对医疗保健行业产生重大影响，引发临床工作、客户服务和预测分析等领域的变革。③在爱思唯尔，我们的技术人员将人工智能机器学习算法应用到大量数据中，使从用户那里“学习”的自适应技术能够提供个性化的体验。④例如，在护理和医疗辅助教育中，Sherpath 通过评估和模拟跟踪学生的互动，根据他们的理解能力和学习风格定制个人体验。

本段一共四句话。第一句的主语是“人工智能”，谓语是系动词“是”，后接宾语“关键角色”，前半句为条件状语，可以采用介词短语“from...to + 名词词组”的结构进行翻译。因此，本句可以处理为“状语 + 主系表”结构：“From wearable technology for patients to technology that assists doctors with surgery, artificial intelligence is already a key player in medicine.”。

第二句的主干部分为“这一领域产生影响”。“随着……”作为句子的状语，在翻译时可以用介词引导，并用非谓语动词对“预计”“引发变革”进行翻译。第二句可以译为：“With overall AI market growth projected to reach $16.6 billion over the next six years at a compound annual growth rate of 63%, this sector is having a major impact on the healthcare industry, sparking changes in clinical work, customer service, predictive analytics, and other areas.”。

第三句的句子结构比较简单，依旧可以用非谓语动词作状语。因此，本句可以译为：“At Elsevier, our technologists apply AI machine learning algorithms to vast amounts of data, enabling adaptive technologies that ‘learn’ from the user to provide a personalized experience.”。**其中第一个分词 learning 是定语，第二个分词 enabling 是状语。**

第四句的主干部分为“Sherpath 跟踪学生的互动”，其他都是补充说明成分，可以译为介词短语和分词形式。整句可以译为：“For example, in nursing and medical assisting education, Sherpath tracks students’ interactions through assessments and simulations to customize their experience based on their understanding and learning style.”。

IV. Translate the following paragraph into English.

将人工智能引入广泛的健康服务中，使医务人员专注于照顾患者个人，这一趋势是不可避免的。随着人工智能技术的日益成熟，医务人员将会越来越依赖它。但是，在可预见的未来，人类在复杂的医疗问题上永远拥有最终发言权。因此，对于人工智能研发公司来说，重要的是传达出医务人员的角色演变以及机器不断增长的能力这一信息。我们需要预见患者日益关注的问题和公众利益，以便在一个更多由机器主导的医疗环境中管理这些预期。

Workshop

I. Choose three applications of artificial intelligence in facial recognition and do research on their working theories. Then make a five-minute oral presentation to the class.

No.	Applications of AI	Working Theories
1.		
2.		
3.		

II. Work in groups and brainstorm together how facial recognition technology works in each of the following aspects. Each group member is responsible for a report on one aspect. For example, "How does facial recognition work in security?" is a suggested title for the first aspect.

- Security
- Mobile Phones
- Finance
- Shopping

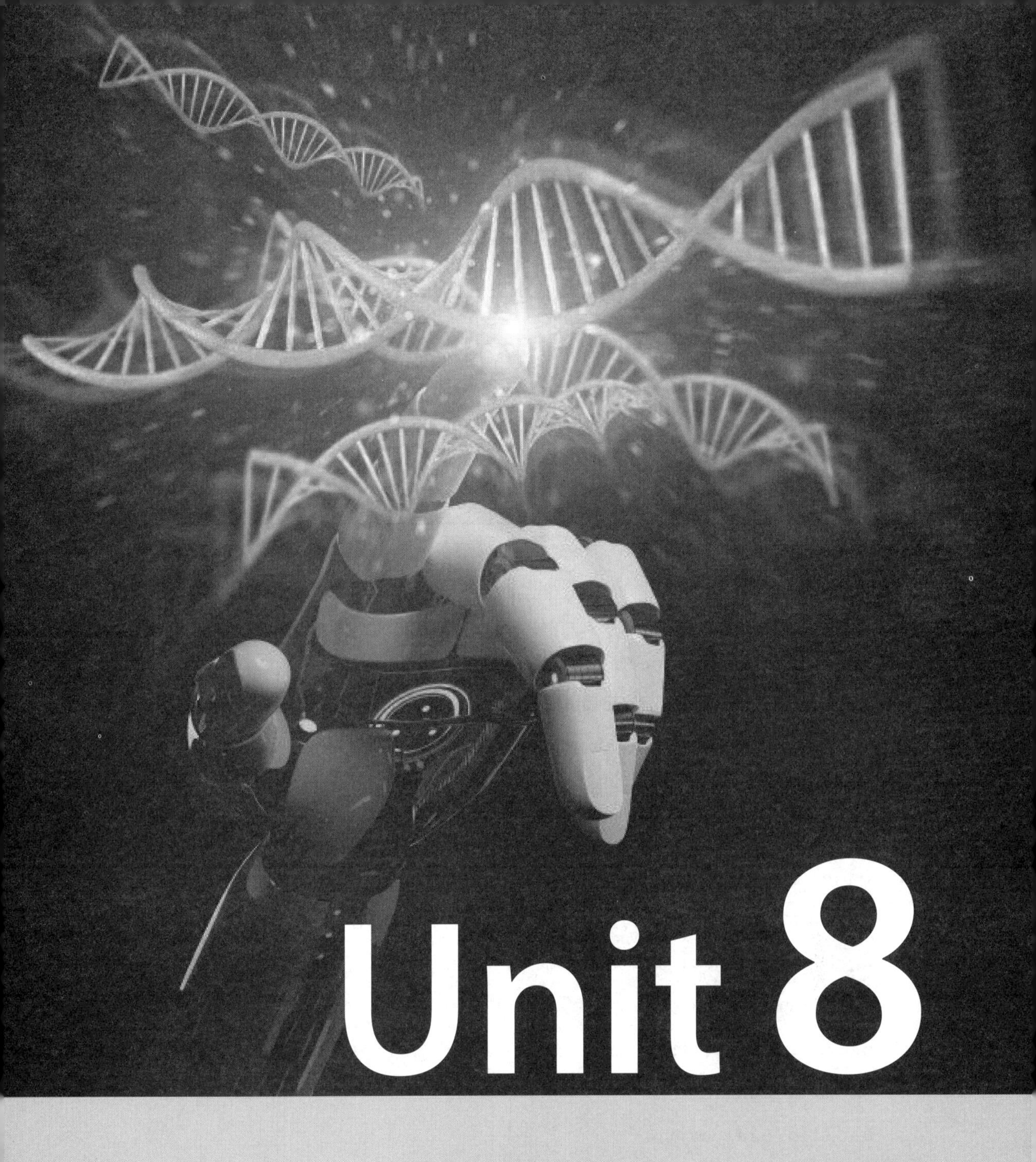

Unit 8

The Future of Artificial Intelligence

Learning Objectives

In this unit, you will learn:

- the risks posed by artificial intelligence in the future;
- the ways to create a promising future of artificial intelligence;
- big strides made by China in artificial intelligence;
- writing skills—stating reasons;
- translating skills—non-finite verbs.

Lead-in

I. Discuss the following questions with your partners.

1. Have you ever imagined the future of artificial intelligence? What will it be like?
2. Do you prefer an AI-driven world or a human-driven world? Why?
3. What can we do to ensure that AI will operate safely when it competes with humans in the future?

II. Do you agree or disagree with the following statements? Give your reasons.

1. Superintelligence by 2100 is inevitable.
2. Superintelligence by 2100 is impossible.
3. AI cannot control humans.
4. AI can control humans.
5. AI will turn evil.
6. AI will turn conscious.

Artificial Intelligence: Our Final Invention?

1 Lately I've **become obsessed with** an issue so daunting that it makes even the biggest "normal" questions of public life seem tiny. I'm talking about the risks posed by "runaway" artificial intelligence. What happens when we share the planet with self-aware, self-improving machines that evolve beyond our ability to control or understand? Are we creating machines that are bound to destroy us?

2 I know when I put it this way, it sounds like science fiction, or the **ravings** of a **crank**. So let me explain how I came to find the issue.

3 A few years ago I read **chunks** of Ray Kurzweil's book *The Singularity Is Near*. Kurzweil argued that what **sets** our age **apart** from all previous ones is the accelerating pace of technological development—an acceleration made possible by the **digitization** of everything. Because of this unprecedented pace of change, he said, we're just a few decades away from basically **meshing** with computers and transcending human biology (think Google, only much better, inside your head). This development will **supercharge** notions of "intelligence", Kurzweil predicted, and even make it possible to upload digitized versions of our brains to the cloud so that some form of "us" lives forever.

4 Mind-blowing and unsettling stuff, to say the least, may exist. If Kurzweil is right, I recall thinking: What should I tell my daughter about how to live—or even about what it means to be a human?

5 Kurzweil has since become **enshrined** as America's uber-optimist on these trends. He and other advocates say accelerating technology will soon equip us to solve our greatest energy, education, health, and climate challenges, extending the human lifespan indefinitely.

6 But a camp of **worrywarts** have **sprung up** as well. The skeptics fear that a toxic mix of artificial intelligence, robotics, and biotechnology and **nanotechnology** could make previous threats of nuclear devastation seem "easy" to manage by comparison. These people aren't cranks. They're folks like Jaan Tallinn, the 41-year-old Estonian programming **whizz** who helped create Skype and now fears he is more likely to die from some advanced AI who **runs amok** than from cancer or heart disease.

7 Now comes James Barrat with a new book—*Our Final Invention: Artificial Intelligence and the End of the Human Era*—that accessibly **chronicles** these risks and how a number of top AI researchers and observers see them. If you read just one book that makes you confront scary high-tech realities that we'll soon have no choice but to address, make it this one.

8 In an interview the other day for my podcast show *This...Is Interesting*, Barrat, a documentary filmmaker, noted that every technology since fire has brought both promise and **peril**. How should we weigh the balance of AI?

9 It turns out that in talking with dozens in the field, Barrat found that everyone is aware of the potential risks of "runaway AI", but no one spends any time on it. Why not? Barrat **surmised** that "normalcy bias"—which holds that if something awful hasn't happened until now, it probably won't happen in the future—**accounts for** the silence.

10 Many AI researchers simply assume we'll be able to build "friendly AI" systems that are programmed with our values and with respect for humans as their creators. When pressed, however, most researchers admit to Barrat that this is wishful thinking.

11 The better question may be this: Once our machines become literally millions or trillions of times smarter than we are (in terms of processing power and the capabilities this enables), what reason is there to think they'll view us any differently than we view ants or pets?

12 The military applications of AI guarantee a new arms race, which the Pentagon and the Chinese are already quietly engaged in. AI's endless commercial applications assure an equally competitive sprint by major firms. IBM, Barrat said, has been **laudably** transparent with its plans to turn its *Jeopardy!*-playing "Watson" into a peerless medical **diagnostician**. But Google—which hired Kurzweil earlier this year as director of engineering, and which also has a former head of the Pentagon's advanced research agency **on the payroll**—isn't talking.

13 Meanwhile, the military is already debating the ethical implications of giving autonomous **drones** the authority to use **lethal** force without human intervention. Barrat sees the coming AI crisis as **analogous** to nuclear fission, which inspired passionate debates over how to pursue these technologies responsibly.

14 At the end of our interview, I asked Barrat what I thought was a joke. "I know you've got a **grim** view of what may lie ahead," I said. "But does that mean you're buying property for your family on a desert island just in case?"

15 "I don't want to really scare you," he said, after half a chuckle. "But it was alarming how many people who I talked to and who are highly placed in AI have retreats that are sort of '**bug-out**' houses to which they could flee if it all hits the fan."

16 Whoa.

17 It's time to take this conversation beyond a few hundred technology sector insiders or even those reached by Barrat's indispensable wake-up call. In his State of the Union Address next month, President Obama should set up a presidential commission on the promise and perils of artificial intelligence to **kick-start** the national debate AI's future demands.

Notes

Ray Kurzweil an American author, computer scientist, inventor, and futurist. He is involved in fields such as optical character recognition (OCR), text-to-speech synthesis, speech recognition technology, and electronic keyboard instruments. He has written books on health, artificial intelligence, transhumanism, the technological singularity, and futurism.

The Singularity Is Near: When Humans Transcend Biology

a 2005 non-fiction book about artificial intelligence and the future of humanity by inventor and futurist Ray Kurzweil. In it, Kurzweil describes his law of accelerating returns which predicts an exponential increase in technologies like computers, genetics, nanotechnology, robotics, and artificial intelligence. Once the singularity has been reached, Kurzweil says that machine intelligence will be infinitely more powerful than all human intelligence combined.

Jaan Tallinn an Estonian programmer, investor, and physicist, who participated in the development of Skype in 2002 and Kazaa, a file-sharing application, in 2000.

James Barrat an American documentary filmmaker, speaker, and author of the non-fiction book *Our Final Invention: Artificial Intelligence and the End of the Human Era.*

Our Final Invention: Artificial Intelligence and the End of the Human Era

a 2013 non-fiction book by the American author James Barrat. The book discusses the potential benefits and possible risks of human-level or super-human artificial intelligence. One of those supposed risks is the extermination of the human race.

Words and Expressions

become obsessed with to be exclusively or solely focused on somebody/something 沉迷于，着迷于

ravings /ˈreɪvɪŋz/ *n.* [pl.] crazy statements that have no meaning 胡言乱语，疯话

crank /kræŋk/ *n.* a person who has strange or unusual ideas 古怪的人，怪人

chunk	/tʃʌŋk/	*n.*	a part of something, especially a large part 相当大的部分
set…apart			to make somebody/something different from or better than others 使与众不同；使突出
digitization	/ˌdɪdʒɪtaɪˈzeɪʃən/	*n.*	the process of changing data into a digital form that can be easily read and understood by a computer 数字化
mesh	/meʃ/	*v.*	to fit together or match closely, especially in a way that works well 紧密配合；相互协调
supercharge	/ˈsuːpətʃɑːdʒ/	*v.*	to make something stronger, more powerful, or more effective 使增强；使更有效
enshrine	/ɪnˈʃraɪn/	*v.*	to preserve or cherish as scared 把……奉为神圣，使……神圣不可侵犯
worrywart	/ˈwʌriwɔːt/	*n.*	a person who often worries, especially about things that are not important 杞人忧天的人，自寻烦恼的人
spring up			to suddenly appear or begin to exist 突然冒出，涌出
nanotechnology	/nænəʊtekˈnɒlədʒi/	*n.*	the branch of technology that deals with structures that are less than 100 nanometers long 纳米技术
whizz	/wɪz/	*n.*	a person who is very good at something 能手；奇手；高手
run amok			to be out of control and act in a wild or dangerous manner, especially in a public place 狂暴，发狂
chronicle	/ˈkrɒnɪkl/	*v.*	to make a record or give details of something in the order of happening 记述，记录（大事）；把……载入编年史
peril	/ˈperəl/	*n.*	the fact of something being dangerous or harmful 祸害；险情
surmise	/səˈmaɪz/	*v.*	to guess something, without having much or any proof 推测，猜测，臆测
account for			to give reasons for something 说明（原因、理由等），解释
laudably	/ˈlɔːdəbli/	*adv.*	in a way that deserves to be praised or admired 值得赞扬地

diagnostician	/ˌdaɪəgnɒsˈtɪʃn/	*n.*	a specialist or an expert in making diagnoses 诊断专家
on the payroll			employed by a particular company 被雇佣的；工资单上的
drone	/drəʊn/	*n.*	an aircraft that does not have a pilot but is controlled from the ground, especially used for dropping bombs or for surveillance 无人驾驶飞机
lethal	/ˈliːθl/	*adj.*	able to cause death 致命的，可致死的
analogous	/əˈnæləgəs/	*adj.*	similar in some way to another thing or situation and therefore able to be compared with it 相似的，类似的
grim	/grɪm/	*adj.*	unpleasant and depressing 令人担忧的，令人沮丧的
bug out			to depart hurriedly 撤退；散逃；匆忙离开
kick-start	/ˈkɪk stɑːt/	*v.*	to do something to help a process or project start more quickly 促使……开始

Useful Term

normalcy bias 正常化偏见

Exercises

Comprehension Check

I. Answer the following questions according to the text.

1. What are the benefits of artificial intelligence according to the advocates?
2. What are the risks of artificial intelligence according to the worrywarts?
3. According to James Barrat, why is there no one spending time on the potential risks of "runaway AI"?
4. How should we weigh the balance of artificial intelligence?
5. What are the applications of artificial intelligence mentioned by the author?

II. Read the following statements carefully and decide whether they are true (T) or false (F) without turning back to check the text.

1. ______________ When the author uses the word "runaway", he refers to the risks posed by artificial intelligence.
2. ______________ According to Ray Kurzweil, what sets our age apart from all previous ones is the emergence of artificial intelligence.
3. ______________ Ray Kurzweil is one of those pessimists concerning the future of artificial intelligence.
4. ______________ Jaan Tallinn is worried that he may die from some advanced AI.
5. ______________ Many AI researchers simply suppose we'll have the ability to build "friendly AI".

III. Choose the best answer to each of the following questions according to the text.

1. In Ray Kurzweil's view, our age is different from all the previous ones in that ______________.
   A. there is an increasing number of "runaway" technologies in our age
   B. there is a rocketing pace of development of technologies in our age
   C. there are more and more artificial intelligence researchers in our age
   D. there are more and more climate changes in our age
2. Ray Kurzweil and other advocates are optimistic about artificial intelligence because they believe the technology will ______________.
   A. help us solve various problems in society
   B. be controlled by human beings
   C. develop at an accelerating pace
   D. evolve beyond our ability to control or understand
3. IBM has a plan to develop its supercomputer "Watson" into ______________.
   A. a matchless expert in biology
   B. an unparalleled expert in healthcare
   C. an incomparable expert in education
   D. an unrivalled expert in medicine
4. What is the attitude of James Barrat towards artificial intelligence?
   A. Optimistic. B. Pessimistic.
   C. Indifferent. D. Neutral.

5. The purpose of the text is to ______________.

   A. convince people of the benefits of artificial intelligence

   B. arouse people's interests in artificial intelligence

   C. remind people of the risks of artificial intelligence

   D. encourage people to develop artificial intelligence technology

Vocabulary Building

IV. Fill in the following blanks with the words and phrases given in the box. Change the form if necessary.

mesh	become obsessed with	evolve	peril	chronicle
account for	laudably	surmise	transcend	grim

1. Tom has ______________ AI technology for as long as he can remember—starting with the first time he ever glimpsed a robot.
2. With no news from the explorers, we can only ______________ their present position.
3. The study's authors say it's the first of its kind, and ______________, it was published on an open-access platform, so you can read it in full here.
4. These headphones ______________ the boundaries of consumer use as an ideal solution for the casual listener or the audiophile.
5. The budget plan offered a(n) ______________ view of the city's economy, saying there was little hope for a major expansion.
6. Is nuclear proliferation the greatest ______________ now facing the world?
7. Our future plans have to ______________ with the present practices.
8. As medical knowledge and technology continue to ______________, it has become possible to perform certain surgeries in the hospital.
9. The book ______________ the writer's coming to natural term with his illness.
10. The peasants were worried because the late frosts ______________ the poor fruit crop this year.

V. Match the words in the left column with the explanations in the right column.

1. upload	A. a person who doubts the truth or value of an idea or belief
2. daunting	B. the process of dividing the nucleus of an atom, resulting in

the release of a large amount of energy

3. skeptic　　C. the increase in the speed of something, or its ability to go faster

4. fission　　D. to copy and move programs or information to a larger computer system or to the Internet

5. acceleration　　E. making you feel slightly frightened or worried about your ability to achieve something

Word Formation

VI. Fill in the following blanks with the words in capitals. Change the form if necessary. An example has been given.

e.g. *It accessibly chronicles these risks and how a number of top AI researchers and observers see them.* RESEARCH

1. A good ______________ would not lead the candidate to the desirable answer. INTERVIEW
2. She was elected as ______________ of the campaign group. LEAD
3. I would prefer not to leave this job to John while he is still a ______________. BEGIN
4. ______________ can break their journey in Singapore if they wish. TRAVEL
5. A convicted ______________ was executed in North Carolina yesterday. MURDER

Translation

VII. Translate the following sentences into English with the words and phrases in brackets.

1. 加速发展的人工智能技术能很快帮我们解决教育、能源和健康方面最大的问题，从而无限期延长人类寿命。(equip to; extend; indefinitely)

____

____

2. 许多人工智能研究员想当然地认为，我们能够创建“友好的人工智能”系统，这些系统会采用我们的价值观，并尊重它们的创造者——人类。(assume; be programmed with)

____

____

3. 人工智能的商业应用是无限的，这使大公司之间的竞争进入白热化阶段。(commercial; intensify)

____

____

Text B

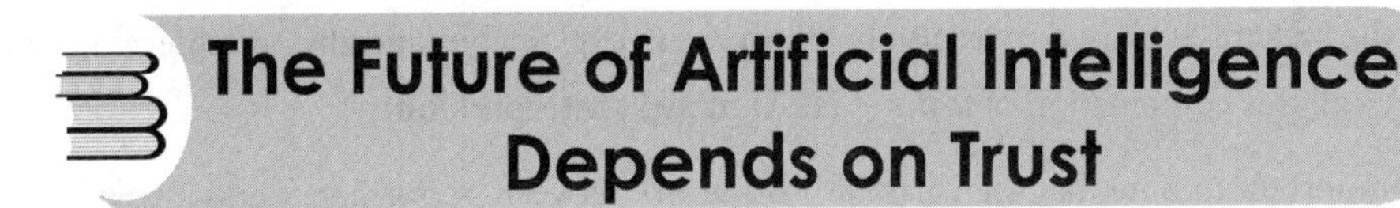

The Future of Artificial Intelligence Depends on Trust

1 Purchasing a home or car is a really exciting moment in one's life. Consumers may be comfortable with and even appreciate **data-driven** recommendations in the search process, for example, from websites that suggest homes based on properties they've previously viewed. But what if the decision to grant a **mortgage** or auto loan is made by a machine-learning algorithm? And what if the logic behind that algorithm's decision, especially if it rejects the application, is unclear? It's hard enough being denied a loan after going through the traditional process; being **turned down** by an AI-powered system that can't be explained is much worse. Consumers are left with no way to know how to improve their chance of success in the future.

2 Elsewhere, for patients and their doctors, the promise of AI programs that can detect signs of disease at ever-earlier stages is the cause for celebration. But it can also be the cause for **consternation**. When it comes to medical diagnosis, the **stakes** are exceedingly high; a **misdiagnosis** could lead to unnecessary and risky surgery or to the **deterioration** of the patient's health. Physicians must trust the AI system in order to confidently use it as a diagnostic tool, and patients must also trust the system if they are to have confidence in its diagnosis.

3 As an increasing number of companies in a range of industries adopt machine learning and more advanced AI algorithms, such as deep neural networks, their ability to provide understandable explanations for all the different stakeholders becomes critical. Yet some machine-learning models that **underlie** AI applications qualify as black boxes, meaning we can't always understand exactly how a given algorithm has decided what action to take. It is human nature to distrust what we don't understand, and much about AI may not be completely clear. And since distrust **goes**

hand in hand with lack of acceptance, it becomes **imperative** for companies to open the black box.

4 Deep neural networks are complicated algorithms modelled after the human brain, designed to recognize patterns by grouping raw data into **discrete** mathematical components known as **vectors**. In the case of medical diagnosis, this raw data could come from patient imaging. For a bank loan, the raw data would be made up of payment history, **defaulted** loans, credit score, perhaps some **demographic** information, other risk estimates, and so on. The system then learns by processing all the data, and each layer of the deep neural network learns to recognize progressively more complex features. With sufficient training, the AI may become highly accurate. But its decision processes are not always transparent.

5 To open up the AI black box and facilitate trust, companies must develop AI systems that perform reliably—that is, make correct decisions—time after time. The machine-learning models on which the systems are based must also be transparent, explainable, and able to achieve repeatable results. We call this combination of features an AI model's **interpretability**.

6 It is important to note that there can be a **trade-off** between performance and interpretability. For example, a simpler model may be easier to understand, but it won't be able to process complex data or relationships. Getting this trade-off right is primarily the domain of developers and analysts. But business leaders should have a basic understanding of what determines whether a model is interpretable, as this is a key factor in determining an AI system's **legitimacy** in the eyes of the business' employees and customers.

7 Data integrity and the possibility of unintentional biases are also concerns when integrating AI. In a 2017 PwC CEO Pulse survey, 76% of respondents said potential for biases and lack of transparency were **impeding** AI adoption in their enterprises; 73% said the same about the need to ensure governance and rules to control AI. Consider the example of the AI-powered mortgage loan application evaluation system. What if it started denying applications from a certain demographic because of human or systemic biases in the data? Or imagine if an airport security system's AI program **singled out** certain individuals for additional screening at airport security on the basis of their race or ethnicity.

8 Business leaders faced with ensuring interpretability, consistent performance, and data integrity will have to work closely with their organizations' developers and analysts. Developers are responsible for building the machine-learning models, selecting the algorithms used for the AI application, and verifying that the AI was built correctly and continues to perform as expected. Analysts are responsible for validating the AI model created by the developers to address the business need at hand. Finally, management is responsible for the decision to deploy that the system, and must be prepared to take responsibility for the business impact.

Note

PwC PricewaterhouseCoopers (doing business as PwC), a multinational professional services network headquartered in London, the United Kingdom. It is the second largest professional services firm in the world, and is one of the Big Four auditors, along with Deloitte, EY, and KPMG.

Words and Expressions

data-driven /ˈdeɪtə drɪvn/ *adj.* based on or decided by collecting and analysing data 数据驱动的

mortgage /ˈmɔːɡɪdʒ/ *n.* an agreement that allows you to borrow money from a bank or similar organization, especially in order to buy a house; the amount of money itself （尤指购房的）按揭；抵押贷款

turn down to decline to accept 拒绝，驳回

consternation /ˌkɒnstəˈneɪʃn/ *n.* a sad, worried feeling after you have received an unpleasant surprise 惊愕，惊恐，惊慌失措

stake /steɪk/ *n.* something that you risk losing, especially money, when you try to predict the result of a race, game, etc., or when you are involved in an activity that can succeed or fail 赌注，赌金

misdiagnosis /ˌmɪsdaɪəɡˈnəʊsɪs/ *n.* a decision that a person has a particular illness or condition when in fact he has a different one 错误的诊断

deterioration /dɪˌtɪəriəˈreɪʃn/ *n.* the state of becoming worse 恶化；退化

underlie /ˌʌndəˈlaɪ/ *v.* to be the basis or cause of something 构成……的基础；作为……的原因

go hand in hand with to go together with 与……共同行动；与……相配合；与……一致

imperative /ɪmˈperətɪv/ *adj.* extremely important and needing immediate attention or action 重要紧急的；迫切的；急需处理的

discrete /dɪˈskriːt/ *adj.* independent of other things of the same type 分离的；各自的，单独的

vector /ˈvektə(r)/ *n.* something physical such as a force that has size and direction 矢量，向量

default /dɪˈfɔːlt/ *v.* to fail to do something, such as pay a debt, that you

			legally have to do 拖欠，不履行债务；违约
demographic	/ˌdeməˈɡræfɪk/	*adj.*	relating to demography (= the study of populations and different groups that make them up) 人口统计的，人口学的
interpretability	/ɪnˌtɜːprɪˈtəbɪlɪti/	*n.*	the quality that something can be explained 可解释性
trade-off	/ˈtreɪd ɔf/	*n.*	a situation in which you balance two opposing situations or qualities 权衡；协调
legitimacy	/lɪˈdʒɪtɪməsi/	*n.*	the quality of being based on a fair or acceptable reason 合理性；合法性
impede	/ɪmˈpiːd/	*v.*	to delay or stop the progress of something 阻碍，阻止
single out			to select from a group 挑选，选出

Useful Term

deep neural network	深度神经网络

Exercises

Comprehension Check

I. Identify the paragraph from which the information is derived.

1. ______________ To open up the AI black box, companies should develop AI systems that perform reliably from time to time.
2. ______________ It is human nature to distrust what we don't understand.
3. ______________ Developers are responsible for checking that the AI is built in a correct way and continues to perform as expected.
4. ______________ It is very important to get a trade-off between the performance and interpretability of an AI model.
5. ______________ A misdiagnosis can result in unnecessary and risky surgery or even the exacerbation of the patient's disease.

6. ______________ Deep neural networks are designed to recognize patterns by dividing raw data into groups of discrete mathematical components.
7. ______________ When integrating AI, we should also be concerned about data integrity and the possibility of unintentional biases.
8. ______________ If the decision to grant a mortgage is made by a machine-learning algorithm, it may cause problems to people.
9. ______________ According to a survey, most of the respondents believed possible biases and lack of transparency prevented the use of AI in enterprises.
10. ______________ Business leaders should have a basic understanding of what decides whether a model can be explained.

II. Answer the following questions according to the text.

1. What is the meaning of some machine-learning models qualifying as black boxes?
2. How do the deep neural networks help deal with the raw data?
3. How can a trade-off between performance and interpretability be obtained?
4. What are the possible concerns when integrating AI?
5. What are the different roles of developers and analysts?

Vocabulary Building

III. Fill in the following blanks with the words and phrases given in the box. Change the form if necessary.

default	single out	imperative	turn down	stake
impede	go hand in hand with	mortgage	underlie	discrete

1. People who ______________ on their mortgage repayments may have their home repossessed.
2. Most developers reserve the right to ______________ a property they think is virtually unsaleable.
3. These small companies now have their own ______________ identity.
4. Poverty tends to ______________ disease, and raising people's incomes usually helps to improve their health.
5. Federal employees, however, may not be immunized from taxes, as the tax wouldn't in any way ______________ government activities.

6. Given the high ____________ for both hardware makers and software suppliers, neither side is likely to give up easily.

7. After falling behind with his ____________ repayments, he now faces eviction from his home.

8. You can't just ____________ young people when you talk about what's wrong with the society and country.

9. The concern of psychology as a basic science is in understanding the laws and processes that ____________ behavior, cognition, and emotion.

10. I feel it is ____________ to guarantee transparency with regard to the financial mechanisms for allocating funds.

Text C

China Makes Big Strides in Artificial Intelligence

1 The World Artificial Intelligence Conference based on the theme of "Intelligent Connectivity, **Infinite** Possibilities" was held on August 29–31, 2019 in Shanghai. The conference attracted both domestic and international AI scientists, entrepreneurs, government leaders, and officials to discuss the latest industry trends and technological advancements. An accompanying exhibition covering an area of almost 15,000 square meters displayed **cutting-edge** technologies related to industrial ecology, AI urban application, autonomous vehicle operation, etc. Nearly 400 home-grown as well as global companies participated in the event.

adj. 无限的，无穷尽的

adj. 尖端的，领先的

2 **Strikingly**, after functioning as a "factory to the world" for almost four decades, China is now working **strenuously** towards becoming an innovation hub for cutting-edge and sophisticated technologies such as big data, virtual reality, and AI.

adv. 引人注目地

n. 全力以赴地

3 A recent report by the Chinese Institute of New Generation Artificial Intelligence Development Plan recognized the rapid strides being made by the country in recent years and **highlighted** its massive potential in transforming the future shape of the Chinese economy. Also, according to a recent study by PwC, the global GDP could expand by 14% by 2030 due to the application of AI. The study further emphasized that China is expected to contribute almost $7 trillion of the $15.7 trillion of the global wealth generated in this way in the period, compared with American contribution of $3.7 trillion.

v. 突出，强调

4 This ambition to **ascend** the global value chain became quite distinct in 2017 when China **stepped up** its efforts to build itself into a strong country with advanced manufacturing, pushing for a deeper integration between the real economy and foremost technologies including the Internet, big data, and artificial intelligence. In the same year, the Ministry of Industry and Information Technology revealed a three-year plan to facilitate the evolution of AI and machine learning up to 2020.

v. 上升，提升

增加，加大

5 China's overall ambition can be **broken down into** three parts: maintaining pace with AI technologies by 2020, accomplishing AI breakthroughs by 2025, and becoming a world leader in AI by 2030. To achieve this, it has been pouring vast amounts of money into AI-based research and development. In 2017, Chinese venture-capital investors contributed almost 48% of the entire AI venture funding worldwide. During the same year, Chinese start-ups raised $4.9 billion in comparison with the $4.4 billion achieved by their North American counterparts.

分成，分解成

6 China's big tech companies such as Alibaba, Baidu, and Tencent have also been investing billions of dollars into AI-based research and development. For instance, in March 2017, Baidu established the National Engineering Laboratory of Deep Learning Technology. The primary ambition of this facility is to conduct research in image and voice recognition, new types of human-machine interaction, and **biometric** identification. Likewise, e-commerce giant Alibaba pledged to invest $15 billion annually up to 2020 in AI-related technologies. The company has already introduced AI cloud services to the healthcare and manufacturing industries.

adj. 生物统计的

7 These initiatives, government's regulatory support, availability of a

growing AI **talent pool**, combined with large amounts of data generated by multiple apps, are helping the country to make considerable strides in this field. Local governments are also backing the AI industry in both finance and resources. Shanghai has recently launched several policies with the intent to establish itself as a base for high-quality AI development. The city has begun constructing the country's first AI innovation application **forerunner** area along with the national new-generation AI innovation and development zone. The latter will work towards **bolstering** the innovation capability, enhancing the environment of **entrepreneurship**, and thus becoming a driver of AI-based research and innovation. Already, Shanghai is home to almost 1,000 AI-related companies, including global giants such as Microsoft, Amazon, and SAP, who have established their AI research institutes in the city. It aims to become a global AI center by further expanding the industry to more than 100 billion yuan ($14.4 billion) by 2020, according to the plan launched by Shanghai municipal authorities last year.

人才库
n. 先驱，先行者
v. 加强；提高
n. 企业家精神

8 Meanwhile, last November, Beijing constructed a new R & D institution named the Beijing Academy of Artificial Intelligence (BAAI). Besides, it also launched a plan entitled "Zhiyuan Action Plan". Both initiatives intended to support scientists and researchers to make disruptive breakthroughs in AI technologies and systems. The city allocated a fund of $2.1 billion last year for constructing a science and technology park focusing primarily on research and development associated with AI. The park, hoping to become home to more than 400 enterprises eventually, is expected to be completed in the next five years.

9 Kai-Fu Lee, author of the *AI Superpowers*, says: "In the age of AI, data is the new oil. As AI begins to '**electrify**' new industries, China's embrace of the messy details of the real world will give it an edge on Silicon Valley."

v. 使激动，使兴奋

10 The world is **on the cusp of** a huge change, and massive diffusion of digital technologies will not just deepen but also transform it fundamentally. Looking at the profound impact of these technologies on authorities, enterprises, and people, it seems that they have no alternative but to join the **bandwagon** to facilitate their development and deploy these technologies fruitfully.

将要进入特定状态
n. 流行，时尚

Exercises

Comprehension Check

I. Answer the following questions according to the text.

1. As a "factory to the world", what is China working hard to become now?
2. Why did the Ministry of Industry and Information Technology reveal a three-year plan in 2017?
3. How many parts can China's overall ambition be divided into? And what are they?
4. What steps are taken to help China make considerable strides in the field of AI?
5. Why are there many policies related to AI development launched in Shanghai?

II. Read the following statements carefully and decide whether they are true (T) or false (F) without turning back to check the text.

1. ______________ According to the study by PwC, America is expected to contribute more in the field of AI to the global wealth than China.
2. ______________ China's aspiration to elevate its global value chain became clear in 2019.
3. ______________ E-commerce giant Alibaba has provided AI cloud services for the healthcare and manufacturing industries.
4. ______________ China's first AI innovation application forerunner area was built in Beijing.
5. ______________ Kai-Fu Lee compared data to the new oil in the age of AI.

III. Discuss the following questions based on your understanding of the future of artificial intelligence.

1. Do you know any polices related to the development of AI in China?
2. If China is to have global influence in the field of AI, is it important to implement proper governance? Why?
3. China's artificial intelligence research is growing in quality, but this field still plays catch-up to some developed countries. What aspect do you think still falls behind and leaves much room for improvement?

Writing Skills

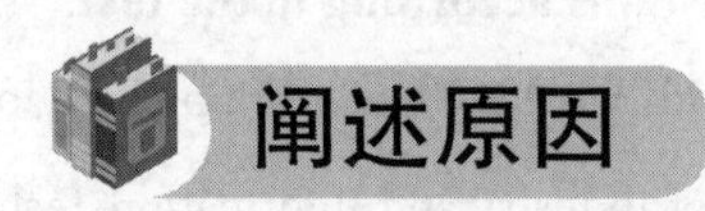

阐述原因

阐述原因（stating reasons）是指对某一现象或观点进行原因分析。英语写作中通常用表因果的连词和序列词来引导句子，如 because、since、in that、firstly、secondly、moreover、furthermore 等。另外还有一些常用的短语和句型，例如：

- be the main/major factor/reason/cause for...
- cause / contribute to / lead to / trigger / bring about / result in / give rise to / generate / account for...
- stem/result from...
- be the result/consequence of...
- be caused by...
- be in part / partly attributed to / be accountable for...
- be a second reason/factor
- be regarded/seen as a root cause
- due to / owing to / because of...
- affect/influence in a positive/negative way
- The reasons are as follows. For one thing...For another...
- There're several factors accounting for / responsible for..., but the following are the typical ones.
- We may blame..., but the real causes are...
- Part of the explanations for it is that...
- One of the most common factors/causes is that...
- Another contributing factor/cause is...
- Perhaps the primary factor is that...
- But the fundamental cause is that...
- Another basic/primary/root reason/cause why...is that...

Choose one argument from the following two and write a short paragraph to state your reasons.

- An AI-driven world is better than a human-driven world.
- A human-driven world is better than an AI-driven world.

____

____

____

____

____

____

Translating Skills

非谓语形式

状语是修饰动词、形容词、副词、各类短语或整个小句的语法成分。按其用途，状语可以分为：时间状语、地点状语、方式状语、原因状语、结果状语、目的状语、条件状语、让步状语、程度状语、伴随状语等。

在翻译科技类文章时，可以灵活处理状语部分的英汉互译。因为这类文章通常需要对某个事物或者各个事物之间的关系进行说明，或者对机器、产品、加工手段、工艺流程和操作方法等进行阐释。为了完整、准确地表达某一概念或介绍某一事物，陈述内容应客观、直叙、简练、准确，避免使用太复杂的句子结构，并注意使用合适的动词时态，使句子既不累赘又语意明确。

动词的非谓语形式（non-finite verbs）不受主语的人称和数的限制。除了不能单独作谓语外，它可以替代句子中的名词性从句、定语从句或状语从句，充当其他句子成分。因此，在翻译时可以使用动词的非谓语形式来处理状语部分，使句子变得简洁、紧凑。

例 1：在整合人工智能时，数据完整性和无意偏差的可能性也是一个问题。

Data integrity and the possibility of unintentional biases are also a concern when integrating AI.

例 2：分析师负责验证由开发人员创建的人工智能模型，以满足手头的业务需求。

Analysts are responsible for validating the AI model created by the developers to address the business need at hand.

例 3：该公司表示，其最终目标是安全地开发人工智能，并广泛分享其影响力，以造福人类。

The company has expressed that its ultimate goal is to develop AI safely and share its reach widely to promote benefits to humanity.

Translation at Sentence Level

I. Translate the following sentences into Chinese.

1. For decades—even prior to the inception of the term—AI has aroused both fear and excitement as humanity contemplates creating machines in our image.

____

____

2. China not only has the world's largest population and looks set to become the largest economy, but also wants to lead the world when it comes to artificial intelligence.

____

____

3. In 2017, China laid out a bevy of milestones to reach by 2020, which include making significant contributions to fundamental research, being a favored destination for the world's brightest talents, and having an AI industry that rivals global leaders in the field.

____

____

4. A big advantage for China is the size of its population, which creates a large potential workforce and unique opportunities to train AI systems, including large patient data sets for training software to predict disease.

____

____

5. As we move closer towards becoming a technology-driven society, AI applications will fulfill the promise that computers would make our lives easier.

____

____

II. **Translate the following sentences into English.**

1. 今天，计算机被广泛应用于解决一些数学问题，这些问题与天气预报和卫星送入轨道有关。

____

____

2. 人工智能技术能引导用户使用全球定位系统软件来找出最有效的路线到达目的地。

____

____

3. 人工智能能让用户使用其子系统进行互联网搜索，并通过自动语音检测程序与智能手机进行联系。

____

____

4. 人工智能的其他日常用途包括提供最新的交通信息、面部识别照片软件和实时语言翻译服务。

____

____

5. 最近，人工智能被用于研究药物之间微妙的相互作用，这些相互作用会给患者身体带来严重的副作用。

____

____

Translation at Paragraph Level

English to Chinese Translation

①Artificial intelligence is already a part of everyday life, from wearable health technology like smart watches and voice assistants to self-driving cars and chatbots. ②And this trend is only expected to continue growing. ③In fact, according to the Brookings Institution, in just over a decade, AI will outperform humans in a variety of tasks, from writing essays to retail work. ④The predicted impact of artificial intelligence on people, business, and industry in coming years is profound, and while some may worry about the dangers of over-automation, there are actually many positive social and economic benefits to AI. ⑤If you're considering a career in programming or other computer professions, you'll be in a position to use AI to potentially solve a variety of human problems across sectors, including humanitarian crises, healthcare needs, and quality control in areas such as manufacturing or product development.

本段一共五句话。第一句是一个简单句，主谓表结构较为明确，但是介词短语 from…to…的结构很长，需要在翻译时将其置于主语之前进行列举，主干结构则采取顺译法。第二句也是一个简单句，可以采用合译法与第一句进行合并。但需要注意被动语态的处理方法，可以将其译成汉语的主动句，并保留原句的主语。前两句可以译为："从可穿戴健康技术（如智能手表和语音助手）到自动驾驶汽车和聊天机器人，人工智能已经成为日常生活的一部分，而且这种趋势只会继续增长。"第三句还是一个简单句。主语 AI 之前是由 according to 引导的评注性状语以及时间状语，在翻译时可以采用顺译法。谓语动词 outperform 指"超越，比……出色"，介词短语 from…to…可以译为含有"……的"结构的定语。本句可以译为："实际上，根据布鲁金斯学会的说法，在短短十年内，人工智能将在从撰写论文到零售的各种工作中超越人类。"

第四句是一个由 and 连接的并列句，前一个分句的主语部分很长，在处理时需要将名词 impact 译成"对……的影响"，后一个分句是由 while 引导的表轻微转折的主从句，在翻译时可以采用顺译法。本句可以译为："预计未来几年，人工智能对人类、商业和工业的影响是深远的，尽管有些人可能会担心过度自动化的危害，但实际上人工智能会带来很多积极的社会和经济效益。"最后一句是一个复合句，由 if 引导的条件状语主从句构成，在翻译时可以采用顺译法。主句中的非谓语动词 to solve a variety of human problems 为目的状语，而 including 是分词作状语，表示"包括……"。本句可以译为："如果你正考虑从事编程或其他与计算机相关的职业，那么你会使用人工智能潜在地解决各种跨行业的与人相关的问题，包括人道主义危机、医疗保健需求以及制造业或产品开发等领域的质量控制。"

III. Translate the following paragraph into Chinese.

Although we don't know the exact future, it is quite evident that interacting with AI will soon become an everyday activity. These interactions will clearly help our society evolve, particularly in regards to automated transportation, cyborgs, handling dangerous duties, solving climate change, friendships, and improving the care of our elders. Beyond these six impacts, there are even more ways that AI technology can influence our future, and this very fact has professionals across multiple industries extremely excited for the ever-burgeoning future of artificial intelligence. AI technology will help us not only live happier and healthier lives, but also conserve time, energy, and money.

Chinese to English Translation

①人工智能发展引起的最大担忧之一是，它将使数千个工作岗位不复存在。②由于在本世纪末之前，科技有可能在数学和编程方面胜过人类，再加上其全天候工作的能力，相关领域中的大多数就业需求将会减少。③另一个担忧是网络安全渗透带来的风险。④由于人工智能算法与其他软件在网络攻击方面的脆弱性没有什么不同，这一风险因素仍然存在不利影响。⑤此外，由于人工智能算法通常需要进行高风险决策，如驾驶汽车和控制机器人，如果处理不当或暗示有误，网络攻击会成功地影响人工智能系统，而这一影响可能会产生破坏性的结果。

本段一共五句话。第一句中的主语是"……之一"结构，对应英语的 one of…，谓语后面用一个 that 引导的表语从句来译，因此本句可以译为："One of the biggest concerns arising from the development of AI is that it will eliminate thousands of jobs."。第二句的主干部分为"需求将会减少"，前面的分句为原因状语，可以用状语从句或介词短语进行处理。第二句可以译为："Due to the likelihood of technology outsmarting humans in mathematics and programming before the end of the century, in combination with their ability to work 24/7, the need for most human-held jobs in related fields will diminish."。

第三句译为主系表结构——"担忧是风险","网络安全渗透带来的"可以作为后置定语。因此本句可以译为："Another concern is the risk posed by cybersecurity infiltration."。第四句和第二句的结构类似，主干部分为"风险因素存在不利影响"，前半句表原因，因此仍然可用状语从句或介词短语处理。本句可以译为："Due to AI algorithms being no different from other software in terms of their vulnerability to cyberattacks, this risk factor remains detrimental."。第五句的主干部分为"影响会产生结果"，前面的分句可处理为原因状语和条件状语，因此本句可以译为："Also, because AI algorithms are often required to make high-stakes decisions, such as driving cars and controlling robots, the impact of successful cyberattacks on AI systems could have devastating results if mishandled or insinuated."。

IV. Translate the following paragraph into English.

互联网用户在访问使用聊天机器人的网站时就已经感受到了人工智能的影响。过去，通过预编程，聊天机器人可以为特定查询提供特定答案。今天，人工智能软件允许聊天机器人和虚拟个人助手回答任何问题，并提供准确的答案。这确实令人印象深刻，但它仍有进步的空间。最终，人工智能驱动的设备将分析人们的言语或行动来解释特定的需求，进而提供更有见地的信息。

____

____

____

____

____

____

I. **Work in groups and hold a debate on whether we are creating machines that are destined to destroy us. Decide your answer to the question and give reasons to defend your argument.**

II. **Read the following case and finish the tasks.**

1. What does it mean for a system to be safe? Does it mean the owner doesn't get hurt? Are "injuries" limited to physical ailments? Clarify the meaning of AI safety and write a brief report on it.

2. To ensure the safety of AI, the transparency of the AI algorithm is needed, but it's not the only desirable feature of a reliable AI. Work in groups to figure out other features an AI system should possess to make it safe to humans.

The possibility of creating thinking machines raises a host of ethical issues. These questions both relate to ensuring that such machines do not harm humans and other morally relevant beings, and to the moral status of the machines themselves.

Media reports are full of examples of AI failures, but most of these examples can be attributed to other causes on closer examination. The list below is curated to only mention failures of intended intelligence. Additionally, these examples include only the first occurrence of a particular failure, but the same problems are frequently observed again in later years.

Finally, the list does not include AI failures due to hacking or other intentional causes. Still, the timeline of AI failures has an exponential trend:

- 1959: AI designed to be a General Problem Solver failed to solve real-world problems.
- 1982: Software designed to make discoveries discovered how to cheat instead.
- 1983: Nuclear attack early warning system falsely claimed that an attack is taking place.
- 2010: Complex AI stock trading software caused a trillion dollar flash crash.
- 2011: E-assistant told to "call me an ambulance" began to refer to the user as an ambulance.
- 2013: Object recognition neural networks saw phantom objects in particular noise images.
- 2015: Automated e-mail reply generator created inappropriate responses.
- 2015: A robot for grabbing auto parts grabbed and killed a man.
- 2015: Image tagging software classified black people as gorillas.
- 2015: Adult content filtering software failed to remove inappropriate content.
- 2016: AI designed to predict recidivism acted as a racist.
- 2016: Designed game NPCs unauthorized superweapons.
- 2016: Patrol robot collided with a child.
- 2016: World champion-level Go playing AI lost a game with a hunan.
- 2016: Self driving car had a deadly accident.
- 2016: AI designed to converse with users on Twitter became verbally abusive.